Galaxies and Galactic Structure

Debra Meloy Elmegreen

Maria Mitchell Professor of Astronomy
Vassar College

Prentice Hall, Upper Saddle River, New Jersey 07458

Library of Congress Cataloging-in-Publication Data

ELMEGREEN, DEBRA MELOY

   Galaxies and galactic structure / Debra Meloy Elmegreen

      p.   cm.

   Includes bibliographical references and index.

   ISBN 0-13-779232-8

   1. Galaxies. I. Title.

   QB857.7.E46 1998   97-43244

   523.1'12--DC21   CIP

Executive Editor: **Alison Reeves**
Production Supervision and Interior Design: **Debra A. Wechsler**
Marketing Manager: **Jennifer Welchans**
Art Manager: **Gus Vibal**
Art Editor: **John Christiana**
Art Director: **Jayne Conte**
Cover Designer: **Rosemarie Votta**
Copy Editor: **Margo Quinto**
Manufacturing Manager: **Trudy Pisciotti**
Illustrations: **Douglas & Gayle Ltd.**

Cover Photograph: *A composite color photo of the barred spiral galaxy NGC 1300 made from deep B-band and I-band images reveals a faint outer ring. Image obtained with the Burrell-Schmidt telescope at Kitt Peak National Observatory by D. Elmegreen, B. Elmegreen (IBM T. J. Watson Research Center), and F. Chromey (Vassar), and enhanced using IBM Data Explorer.*

Printed in the United States of America
10 9 8 7 6 5 4 3 2 1

ISBN 0-13-779232-8

Prentice-Hall International (UK) Limited, *London*
Prentice-Hall of Australia Pty. Limited, *Sydney*
Prentice-Hall Canada Inc., *Toronto*
Prentice-Hall Hispanoamericana, S. A., *Mexico*
Prentice-Hall of India Private Limited, *New Delhi*
Prentice-Hall of Japan, Inc., *Tokyo*
Simon & Schuster Asia Pte. Ltd., *Singapore*
Editora Prentice-Hall do Brasil, Ltda., *Rio de Janeiro*

*This book is dedicated to my family
for their constant love and support:
Bruce, Lauren, and Scott Elmegreen
Anne, Thurston, and Linda Meloy*

Contents

Preface

Undergraduate education in astronomy for serious students of the subject requires an introduction to the observations and theory of planets, stars, interstellar matter, galaxies, and cosmology. These fields are interrelated, so many courses necessarily have some overlap in content. This book introduces observational and theoretical perspectives on galaxies, including their formation, structure, evolution, and distribution. Some courses may be designed to cover stars and galaxies or galaxies and cosmology; others may emphasize the Milky Way or general astrophysics. This book is designed for a one-semester course; the order may be varied to suit other syllabi.

This book developed from a sophomore- to junior-level course taught at Vassar College to astronomy and physics majors and interested nonmajors with introductory astronomy, physics, and calculus backgrounds. It is designed to be self-contained. Introductory texts are often cursory by necessity, whereas graduate texts on galaxies and galactic structure assume physics and mathematics preparation and include details that are beyond the scope of intermediate-level work. Because intermediate levels can vary in complexity, this book incorporates and explains the appropriate physics as needed. Concepts are introduced as required, so some basic material appears throughout the book. For example, conversions from angular to linear measurements are described in Chapter 4 when observations of galaxy sizes are first considered. Chapter 3 includes some fundamental astronomical definitions and a review of star properties, and Chapter 6 reviews gas properties that are essential for studying galaxies, in case this material has not been covered in previous courses. Many ideas about stars are developed over several chapters such as mass functions in Chapter 4, when population synthesis is discussed, and Cepheids in Chapter 12, when distance indicators are compared. Appendix 4 includes a brief summary of fluid dynamics to accompany some chapters.

The Milky Way presents a special challenge to astronomy because of our perspective. Chapter 8 is devoted to studies that are unique to the Galaxy, such as kinematic and structural details. Global perspectives of the Milky Way are often best interpreted in the context of other galaxies, so many discussions of our galaxy are also interwoven throughout other chapters where appropriate.

Each chapter summarizes the learning objectives. The chapter "toolbox" highlights the fundamental concepts presented. Equations that are particularly useful to remember are outlined by a box. Exercises are included at the end of each chapter. Some are problems that draw upon equations and material presented in the chapter; others allow the student to learn more about galaxies by accessing the literature and publicly available data, including websites. Constants, astronomical quantities, and conversions are provided in Appendices 1–3.

Although many ideas are presented in the book, in virtually every aspect of galaxy research there is room for more research and greater advances. Lest the subject appear closed, each chapter poses two unsolved problems that represent some of the many avenues beckoning further exploration.

The order of the book is driven by observational considerations, so galaxies are described first and analyzed later. Thus, a general description of galaxies is followed by a detailed examination of their light distributions, and then kinematic data allow an interpretation of the cause of the structure. Later chapters deal with interactions of galaxies and their distribution in the cosmos.

The field of galaxy research is rapidly changing because of new insights provided by the Hubble Space Telescope, new and better ground-based instrumentation, and supercomputers for theoretical modeling. This book attempts to present fundamental views on galaxies and point out new areas of research, but it cannot be completely thorough or up-to-date. The reader is urged to seek out professional articles as well as books and review articles listed in Further Reading at the end of each chapter in order to stay abreast of current research. The references cited in each chapter are included in the Further Reading so that particular topics of interest can be explored in more detail. Useful Websites, with news releases, data, images, and exercises are also included. My website is http://noether.vassar.edu/Elmegreen.html; I will try to post updates on it, and I welcome your comments and suggestions.

I am grateful to the students enrolled in *Astronomy 212: Galaxies and Galactic Structure* at Vassar College over the past five years who served as the inspiration and sounding board for this book. I thank the following reviewers for their many useful suggestions: Harriet Dinerstein, University of Texas–Austin; William Herbst, Wesleyan University; Stephen Gottesman, University of Florida; John Hoessel, University of Wisconsin–Madison; Kenneth Rumstay, Valdosta State University; Ethan Vishniac, University of Texas–Austin; and James C. White II, Middle Tennessee State University. Special thanks go to my husband, Bruce Elmegreen, for his critical reading of many preliminary versions of the manuscript and his detailed advice.

Debra Meloy Elmegreen

Digitized Sky Survey Images

The Digitized Sky Surveys were produced at the Space Telescope Science Institute under U.S. Government grant NAG W-2166. The images of these surveys are based on photographic data obtained using the Oschin Schmidt Telescope on Palomar Mountain and the UK Schmidt Telescope. The plates were processed into the present compressed digital form with the permission of these institutions.

The National Geographic Society—Palomar Observatory Sky Atlas (POSS-I) was made by the California Institute of Technology with grants from the National Geographic Society. The Second Palomar Observatory Sky Survey (POSS-II) was made by the California Institute of Technology with funds from the National Science Foundation, the National Geographic Society, the Sloan Foundation, the Samuel Oschin Foundation, and the Eastman Kodak Corporation. The Oschin Schmidt Telescope is operated by the California Institute of Technology and Palomar Observatory.

The UK Schmidt Telescope was operated by the Royal Observatory Edinburgh, with funding from the UK Science and Engineering Research Council (later the UK Particle Physics and Astronomy Research Council), until June, 1988, and thereafter by the Anglo-Australian Observatory. The blue plates of the southern Sky Atlas and its Equatorial Extension (together known as the SERC-J), as well as the Equatorial Red (ER), and the Second Epoch [red] Survey (SES) were all taken with the UK Schmidt.

These images were produced with support to Space Telescope Science Institute, operated by the Association of Universities for Research in Astronomy, Inc., from Contract No. NAS5-26555 with the National Aeronautics and Space Administration, and are reproduced with permission from AURA/STScI.

CHAPTER 1

Overview of Galaxies and Their Place in the Universe

Chapter Objectives: to describe the study of galaxies

Toolbox:

lookback time

1.1 Galaxy Studies

The Universe is an energetic place. It contains starlight, radio emission, high-energy photons, microwave background emission left over from the Big Bang, magnetic fields, ordinary matter, and dark matter. Although the cosmos is mostly empty space, the occasional lumps of matter and the radiation and gravitational forces that permeate it make a stimulating and interesting environment. Most detectable matter exists as galaxies, which are often clustered together. The boundaries of groups of galaxies occupy about 5% of the volume of space. Individual galaxies comprise gravitationally bound stars and gas of different kinds and different amounts.

Probably most galaxies formed within the first billion years following the Big Bang that began the Universe in which we live, although some galaxies may be forming today out of galaxies colliding with other galaxies. The content, structure, and motions of galaxies and details of how they form and evolve with time and as a function of their environment are the subjects of this book.

In an ideal laboratory, astronomers would study galaxies over long periods of time to see how they change, would probe them to see how they respond, and would crash them together to investigate the effects of interactions. Instead, galaxies must be analyzed remotely by studying their radiation and the responses to their gravitational fields. Although any one galaxy can be observed for

1

only a snapshot of its lifetime, a coherent scenario of galaxy formation and evolution can be formulated by studying a large variety of galaxies. Of course, not all galaxies are representative. Part of the intrigue of galaxy research lies in trying to understand what stages are common to many galaxies and what stages are due to exceptional circumstances.

The study of galaxies is greatly aided by the finite velocity of light: Light requires a time to reach us that is directly related to the distance it travels. In one year, it travels a *light year,* a distance of about 6 trillion miles. Because the distances to different galaxies vary, we view them as they appeared at different times in the past. This phenomenon is known as the *lookback* time. The nearest large galaxy to us is the Andromeda galaxy (see Figure 1.1), at a distance of 2 million light years; consequently, we see it now as it was 2 million years ago. The most distant objects are quasars (the cores of young galaxies) and protogalaxies (galaxies in formation) at distances of some 15 billion light years, so we observe them now as they were 15 billion years ago. Thus, various evolutionary galactic stages can be seen by viewing different galaxies.

1.2 Galaxy Sizes and Compositions

There is a large range of possible masses and sizes for galaxies: Our own Milky Way galaxy contains about 200 billion stars, distributed mostly in a flattened disk about 100,000 light years across. Some giant ellipticals have 10 times as much mass, whereas dwarf galaxies may have less than one-hundredth as much mass; several percent of a galaxy's luminous mass may be in the form of gas. What we

FIGURE 1.1   *Our nearest large neighbor, Andromeda, or M31. (Imaged with the Burrell-Schmidt telescope at KPNO; B band mosaic by F. Chromey and P. Choi, Vassar College.)*

observe in a galaxy depends on the region of the electromagnetic spectrum that is examined. At optical wavelengths, radiation is primarily from stars, clusters of stars, and H II regions, which are the ionized gas around hot young stars. Star masses range from about one-tenth the mass of the Sun to about 40 times its mass. Low mass stars are more commonly formed than are high mass stars; statistically, high mass stars are mostly found in the largest clouds. Young high mass stars radiate most of their light at short wavelengths because of their high temperatures, so they appear very blue. Near-infrared light comes mostly from the more numerous lower mass, longer-lived stars. Sometimes the distributions of these two groups are very different, as we shall see.

High mass stars evolve quickly, on time scales of 10 million to 100 million years. They provide heavily enriched material to the interstellar medium; all elements heavier than iron are produced when a high mass star detonates as a supernova. Supernovae also release oxygen and other heavy elements made in stellar layers prior to the explosion. Subsequent generations of stars, formed out of this processed gas, have significantly high percentages of elements other than hydrogen and helium. Different types of galaxies vary in their present and past star formation rates, so they have different metal contents today.

The presence of gas is inferred from intermixed dust, which obscures and reddens starlight passing through it. In the radio part of the spectrum, gas clouds can be detected directly. Gas exists in the form of atomic and molecular clouds as well as a more diffuse phase between the clouds; clouds have masses of a few hundred to a million times the mass of the Sun. Clouds of atomic hydrogen emit radiation at a wavelength of 21 cm as electrons make low-energy jumps between the two ground-state hyperfine levels of neutral hydrogen. Molecular clouds emit radiation over a range of wavelengths, depending on the molecules and transitions involved. For example, carbon monoxide molecules, which are about 10^{-4} times as abundant as molecular hydrogen, radiate strongly in the millimeter range as a result of rotational transitions.

In the infrared part of the spectrum, the rotational-vibrational transitions of molecular hydrogen are directly detected, as well as the radiation from dust heated to temperatures between a few tens of degrees and 1000 degrees above absolute zero by nearby embedded protostars and high mass stars. There is also a general diffuse background of radiation caused by dilute starlight.

The ultraviolet (uv) part of the spectrum provides a probe of hot young stars whose radiation peaks at uv wavelengths, as well as gas whose uv emission lines are useful as diagnostics of gas density and temperature. In the high-energy region, x-rays and gamma-rays characterize the energetic nonthermal emission processes associated with violently disturbed regions of the interstellar medium, such as accretion disks of black holes, or thermal emission from supernova remnants and other hot regions. Sometimes light is polarized as a result of the alignment of elongated dust grains in the presence of magnetic fields. Figure 1.2 illustrates the types of objects detected at different wavelengths.

A wide variety of instruments and telescopes has been designed to investigate specific regions of the electromagnetic spectrum. Because the atmosphere di-

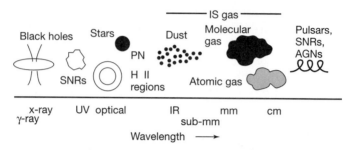

FIGURE 1.2 *Different types of objects radiate in different parts of the electromagnetic spectrum. AGNs, active galactic nuclei; IR, infrared; IS, interstellar; PN, planetary nebulae; SNRs, supernova remnants.*

FIGURE 1.3 *(left) The Multiple Mirror Telescope in Arizona. (From the Harvard public website.) (right) One of the two Gemini 8–m telescopes undergoing construction at Mauna Kea. (From the NOAO public website.)*

FIGURE 1.4 *(left) Hubble Space Telescope (HST) during refurbishing in February 1997. (right) STS–82 shuttle sent to repair the HST. (Images from NASA website.)*

TABLE 1.1 *A sample of telescope facilities for different wavelengths*

Wavelength	Facilities
Gamma ray, x-ray	AXAF (Advanced X-Ray Astrophysics Facility)
	Compton Gamma Ray Observatory
	EGRET (Energetic Gamma Ray Experiment Telescope)
	HEAO-1,2 (High Energy Astrophysical Observatory)
	ROSAT (Roentgen Satellite)
Ultraviolet	EUVE (Extreme Ultraviolet Explorer)
	HST (Hubble Space Telescope)
	IUE (International Ultraviolet Explorer)
Optical	AAT (Anglo-Australian Telescope) (Australia)
	Isaac Newton Telescope, Canary Islands (Spain)
	Carnegie Observatories (Las Campanas, Chile)
	CFHT (Canada-France-Hawaii Telescope) (Hawaii)
	CTIO (Cerro Tololo Inter-American Observatory) (Chile)
	Hobby-Eberly Telescope (Texas)
	HST (Hubble Space Telescope)
	Keck Telescope (Hawaii)
	KPNO (Kitt Peak National Observatory) (Tucson)
	Lowell Observatory (Arizona)
Infrared, submillimeter	IRAS (Infrared Astronomical Satellite)
	SIRTF (Space Infrared Telescope Facility)
	SOFIA (Stratospheric Observatory for Infrared Astronomy), balloons, rockets
	SWAS (Submillimeter Wave Astronomy Satellite)
Millimeter	BIMA (Berkeley-Illinois-Maryland Array) (California)
	FCRAO (Five College Radio Astronomy Observatory) (Massachusetts)
	Nobeyama Radio Observatory (Japan)
	NRAO (National Radio Astronomy Observatory) (W. Virginia)
Centimeter	Arecibo (Cornell University, Puerto Rico)
	Onsala (Sweden)
	Owens Valley (Caltech) (California)
	VLA (Very Large Array, part of NRAO) (New Mexico)
	Westerbork (Germany)

minishes resolution and absorbs infrared and ultraviolet light, high-altitude observatories, satellites, rockets, or balloons are required for many observations. A sample of Earth- and space-based telescopes in use or under construction is shown in Figures 1.3–1.6 and is listed in Table 1.1.

The morphology of a galaxy can be described in terms of the global properties of the radiation (for example, total brightness as a function of wavelength) and in terms of the radiation as a function of position in the galaxy. The latter includes the radial distribution (a function of distance from the galaxy center), the azimuthal distribution (a function of angle around the galaxy at a particular ra-

FIGURE 1.5   *Green Bank Telescope (GBT) under construction in West Virginia. (Image from the NRAO website.)*

FIGURE 1.6   *(left) Berkeley-Illinois-Maryland Array (BIMA) in Hat Creek, California. (Image from the Berkeley website.) (right) Proposed Millimeter Array (MMA.) (Sketch from the NRAO website.)*

dius), and the vertical distribution (a function of distance from the midplane). The motions of luminous matter also provide indirect information about dark matter halos of unknown composition, which may account for more than 90% of a galaxy's total mass.

1.3 Galaxy Shapes

There are many kinds of galaxies, which can be represented well by three basic shapes: elliptical, spiral, and irregular. These galaxies have very different internal motions, types of stars, colors, metallicities, and light distributions. Elliptical galaxies mostly comprise old low mass stars, which are often grouped into globular clusters, and a small percentage of interstellar material, much of which may be highly ionized. Spiral galaxies contain old stars in their central regions and in halos, and they also contain young, high mass stars, often in galactic clusters, that are distributed in a narrow plane called the disk. Irregular galaxies have chaotic shapes, but their stellar properties are similar to those of spirals. Spiral and irregular galaxies are much bluer than ellipticals as a result of ongoing star formation, and a substantial fraction of their luminous mass is in the form of gas. Figure 1.7 shows elliptical and spiral galaxies in the same field of view.

Why do ellipticals differ from spirals and irregulars? The answer may lie partly in the different star formation histories: Ellipticals formed most of their stars in the first billion years after the Big Bang, and less than a percent of their mass is in the form of gas today. Spirals formed some stars right away, but they still have between 5% and 15% of their mass in gas, which can form new stars. Irregulars often have as much as 25% gas. As spinning protogalactic clouds collapse under their own gravity, any gas that has not gone into making stars is deformed to a disk shape, while the already-formed stars remain in eccentric elliptical orbits. Evidently, primordial ellipticals spun more slowly than spirals and irregulars and also may have suffered more interactions, leading to their different star formation rates.

The Milky Way presents a unique challenge for interpretation. On the one hand, regions of star formation can be observed with great clarity because they are so much closer than star formation regions of other galaxies. On the other hand, it is very difficult to discern our spiral structure because the Sun lies in the midplane of the disk. Furthermore, dust obscures most of the optical structure in the disk. We can assemble a coherent picture of our galaxy only by examining other galaxies, for which we have a better perspective. Thus, the study of other galaxies is essential to the study of our own galaxy.

1.4 Galaxy Distributions

Galaxies are clumped together into small groups (such as Stefan's Quintet in Figure 1.7) or larger clusters and superclusters; Figure 1.8 shows a computer simulation of the large-scale structure in our Universe. Our Galaxy is part of the Virgo supercluster, whose existence was first hypothesized in the 1950s. The supercluster is named after the Virgo cluster of galaxies, which comprises nearly 300 galaxies and is the largest cluster in the supercluster. Our own Local Group of galaxies has only about two dozen members, including our large neighboring spiral galaxy, Andromeda, and several dwarf galaxies; the Large and Small Magellanic Clouds are two small galaxies that orbit the Milky Way. There are no large elliptical galaxies in our Local Group, which is partly an effect of our sparse environment.

Environment plays a role in determining the structure of galaxies because of the gravitational perturbations by neighboring galaxies. There are approximately equal numbers of spiral and elliptical galaxies, although they occur more or less frequently in different environments. Dense clusters containing 100 or more members have mostly elliptical galaxies; sparser clusters and small groups of

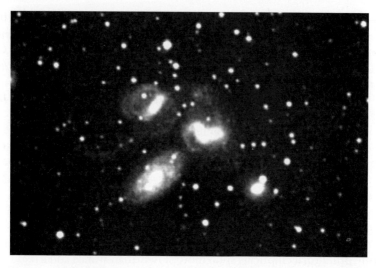

FIGURE 1.7 *Stefan's Quintet is a small group of five galaxies. NGC 7317 in the lower right is an elliptical galaxy; the pair of spirals in the center are NGC 7318a and NGC 7318b; NGC 7319 is in the upper left, and NGC 7320 in the lower left. (Image from the STScI Digital Sky Survey.)*

dozens of galaxies have a mixture of spiral and elliptical galaxies. Irregular galaxies often are dwarfs that tend to orbit larger spirals and ellipticals. A galaxy responding to a tidal encounter can undergo new star formation or alter its shape. Galaxies can form unusually large quantities of stars, form bars, lose material

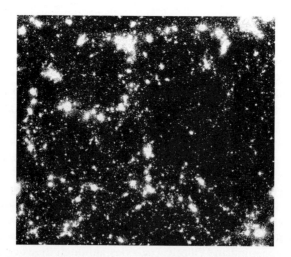

FIGURE 1.8 *Simulation of the large-scale structure of the Universe based on a cold dark matter model. (By J. M. Gelb and E. Bertschinger, Massachusetts Institute of Technology.)*

through tidal disruption, or merge with other galaxies to form different shapes. Sometimes encounters may lead to the formation of separate dwarf galaxies.

We can measure large-scale motions such as the recession of galaxies due to the expansion of the cosmos; these motions are clues to the origin of galaxies as well as to the ultimate fate of the Universe. We can also examine individual galaxy rotation curves, which indicate how objects orbit at different distances from the center of a galaxy, or observe detailed motions of individual objects and how they deviate from the mean motions of other objects. The study of kinematics, or motions of objects, helps in understanding why galaxies have the shapes that they do. The way in which star formation proceeds on global and local scales also depends on these motions. We will consider the details in subsequent chapters.

Useful Websites

The following sites have current news releases and images of general interest:

The National Optical Astronomy Observatories (NOAO) offers updates via http://www.noao.edu

The National Radio Astronomy Observatory (NRAO) site is http://www.nrao.edu

NASA offers info on http://www.nasa.gov; also the latest on shuttle missions via http://www.ksc.nasa.gov/shuttle/missions

The Space Telescope Science Institute (STScI) has Hubble Space Telescope images and information on http://www.stsci.edu

Organizations of professional astronomers include: American Astronomical Society, http://www.aas.org; Astronomical Society of the Pacific, http://www.aspsky.org; International Astronomical Union, http://WWW.lsw.uni-heidelberg.de/iau.html

The Electronic Universe, run by the University of Oregon, has images and simulation movies: http://zebu.uoregon.edu

The magazine *Sky & Telescope* offers http://www.skypub.com

Europe's on-line project is http://www.eso.org/astronomyonline/

Trans-Europe observation homepage is http://www.astro.ku.dk/astronomyonline

The Society for Popular Astronomy in England is http://www.u-net.com/ph/spa

A French base with a large astro server is http://www.univ-rennes1.fr/ASTRO/astro.english.html

A student-run link from the University of Arizona is http://www.seds.org/galaxy/links-astronomy.html

Scientific abstracts and papers can be browsed through the National Extragalactic Database (NED): http//ned.ipac.caltech.edu (or by ftp through the same address); also through http://adsabs.harvard.edu/abstract_service.html

The North American Small Telescope Consortium (NASTeC) website has information on observatories, some of which have observing time available to individuals: http://www.nastec.org (or email to hpreston@valdosta.peachnet.edu)

The Hands-On Universe has images for downloading and observing time available on its member telescopes: http://hou.lbl.gov

Further Reading

Reviews of galaxy research are often found in the following journals:

Astronomy
CCD Astronomy
Discover
Mercury
Scientific American
Sky & Telescope

More technical articles are found in journals such as:

Annual Review of Astronomy and Astrophysics
Astronomical Journal
Astronomy and Astrophysics
Astrophysical Journal
Monthly Notices of the Royal Astronomical Society
Nature
Publications of the Astronomical Society of the Pacific
Science

CHAPTER 2
Galaxy Classification

Chapter Objectives: to describe the basic shapes of galaxies and to present catalogs and atlases

---

Toolbox:

Hubble type   inclination
ellipticity   coordinates
de Vaucouleurs type

2.1 Historical Perspective

Galaxies are vast collections of stars that are held together by mutual gravitational attraction. These stars are generally so far apart that most of a galaxy's volume is empty space, sparsely filled with a dilute hydrogen gas. Star separations compared with their sizes are analogous to grains of sand separated by 10 km. This emptiness means that most galaxies are very faint, because the average light level from all the stars is very low. Only our own Milky Way galaxy, the Large and Small Magellanic Clouds (nearby southern galaxies; see Figure 2.1), and the Andromeda galaxy can be seen readily with the naked eye on a dark night, even though a dozen galaxies have angular diameters larger than that of the full moon. Most other galaxies can be seen only with moderate or large telescopes, which collect much more of their dilute starlight than our eyes alone can.

In 1780, the French astronomer Charles Messier cataloged 103 nebulous objects, many of which turned out to be galaxies. The German-born English astronomer William Herschel, after discovering Uranus in 1781, was named the King's private astronomer in England. He built an 18-in. aperture and then a 4-ft. aperture reflecting telescope. He and his sister Caroline used these telescopes to make a catalog of more than 2000 galaxies. His son John expanded the catalog and compiled it into the *General Catalogue* in 1864. Most of these galaxies appeared only as fuzzy patches of light until 1845, when Lord Rosse of Ireland

11

FIGURE 2.1   *The Large Magellanic Cloud. (From the Electronic Universe website by G. Bothun, University of Oregon.)*

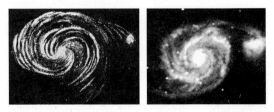

FIGURE 2.2   *Lord Rosse made the first sketch (left) of spiral structure in another galaxy, M51 (***Philosophical Transactions 1850, Vol. 140, p. 499***). Compare this with the B band photo (right) taken with the 1.2-m Schmidt telescope at Palomar Observatory. (By the author.)*

distinguished a spiral shape in the galaxy M51 with his 72-in. reflecting telescope, as shown in Figure 2.2. The term *extragalactic nebulae* is still used today to refer to galaxies, whose true nature as separate systems was not confirmed until the 1920s through the work of Edwin Hubble, V. M. Slipher, Milton Humason, and others.

Galaxies revealed their remarkable internal structure more easily after the advent of astronomical photography in the 1880s, and particularly with the construction of the 100-in. Hooker telescope at Mount Wilson, California, in 1917. Photography allowed exposures perhaps 100,000 times longer than the one-tenth of a second used by the eye to register an image, and the 100-in. diameter telescope could collect about 100,000 times more light per second than the pupil of an eye. This means that the Mount Wilson telescope could register, in a single photograph, 10 billion times more light than the naked eye.

Photographic emulsions have low efficiencies, because only a few incoming photons in a thousand will cause the grains in the emulsions to react and record the light. New technologies such as *charge-coupled devices*, or CCDs, allow greater efficiencies, so more light can be recorded in shorter times. CCDs have *quantum efficiencies* as high as 50%–60%, meaning that every other photon will generate a signal. A CCD instrument is shown in Figure 2.3.

FIGURE 2.3 *2.1-m telescope at KPNO; the CCD camera is in the black box on the bottom. (By the author).*

A photograph or CCD image can show details down to a limiting resolution of about one-half to one second of arc. Resolution is limited by atmospheric turbulence. The CCD scale is 20 times finer than the human eye can resolve. Elimination or reduction of atmospheric seeing effects is possible in space, with high altitude observatories, or with ground-based adaptive optics. Interferometers, which are two or more telescopes coupled together, also increase resolution.

Many of the early observations of galaxies were made at Mount Wilson by the astronomer Edwin Hubble, who realized that there were several basic shapes: elliptical, spiral, lenticular, and irregular. Allan Sandage has spent many years organizing the Hubble collection and expanding it with photographs taken at the 200-in. Hale telescope at Mount Palomar, California, and the 100-in. telescope at Las Campanas, Chile. Morphological studies of galaxies began as visual inspections of photographs, which are sensitive primarily to blue light. Charge-coupled devices are more red-sensitive, and new infrared arrays extend our morphological studies even further. We now know that a galaxy can appear very different in different colors and at different depths of exposure, as we will explore in Chapter 4.

In this chapter, we will review the Hubble classification system and the modifications to it by Gerard de Vaucouleurs. Hubble's galaxy divisions describe the overall galaxy appearance and so are a useful first step in trying to understand galaxies. This classification system has been widely adopted because it turns out to be a division of galaxies according to very different compositions, mass distributions, and kinematics. The attempts to understand all of the ways these basic types of galaxies differ, and the reasons for their underlying differences, constitute a large part of galaxy research.

There are many other classification systems in use besides Hubble's that describe particular features resulting from specific physical events or morphologies in galaxies; these classification schemes will be introduced in subsequent chapters when light distributions are quantified. Thus, Luminosity Class and Yerkes classifications will be covered in Chapter 4 following a discussion of magnitudes, colors, and equivalent spectral types of galaxies, and Arm Classes will be introduced in Chapter 5 together with a discussion of light variations in disks.

2.2 Elliptical Galaxies

Elliptical galaxies have smooth three-dimensional shapes. They appear to us as two-dimensional ovals with central light concentrations and various degrees of apparent flattening, known as the *ellipticity*, ε,

$$\varepsilon = 1 - \frac{b}{a}$$

Here a is the major axis, or apparent longest dimension, and b is the minor axis, or apparent shortest dimension. Hubble denoted elliptical galaxies by the letter E followed by a number, n, ranging from 0 to 7, which describes the ellipticity as a whole number:

$$\boxed{n = 10\left(1 - \frac{b}{a}\right) = 10\varepsilon}$$

With this notation, an E0 galaxy appears round and an E5 galaxy has its major axis twice as long as its minor axis, as shown in Figure 2.4.

Note that this classification describes just the *apparent* ellipticity of a galaxy. A galaxy whose true axes give an ellipticity of 0.5 could appear as any shape from an E0 to an E5, depending on its orientation to our line of sight.

The ellipticity is related to the mathematical definition of *eccentricity*, e, which is given by

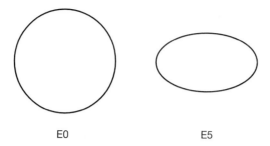

EO   E5

FIGURE 2.4 *The apparent shapes of some elliptical galaxies.*

$$e = \sqrt{1 - \frac{b^2}{a^2}}$$

An E0 galaxy, like a circle, has an eccentricity of 0, whereas an E5 galaxy has an eccentricity of 0.87. Two elliptical galaxies are shown in Figure 2.5.

Elliptical galaxies have more complicated structures than was initially realized. At first they were thought to be either *oblate* like pancakes, in which two axes are equal and longer than the third, or *prolate* like cigars, in which two axes are equal and shorter than the third. Now it is inferred from intensity contours that at least some elliptical galaxies are *triaxial*, meaning that all three axes differ in length. Evidently, there are at least two main types of ellipticals, commonly referred to as *boxy* and *disky*. Boxy ellipticals have a rectangular distribution of light in their centers, whereas disky ones have a slower fall-off of light intensity in their outer than in their inner regions. Some elliptical galaxies also have faint arcs in their outer regions, which may be remnants of mergers with other galaxies. In Chapter 5, the distribution of light in elliptical galaxies will be considered in more detail, and in Chapter 12 interactions will be discussed.

FIGURE 2.5 *NGC 4486 (left) is an E0 galaxy; NGC 3377 (right) is an E5 galaxy. (Images from the STScI Digital Sky Survey.)*

Dwarf ellipticals generally are only one-tenth or one-hundredth the mass of normal ellipticals; they are designated *dE*. Usually they are the companions of larger galaxies.

2.3 Lenticular Galaxies

Some galaxies resemble ellipticals in their central regions but are surrounded by a structureless flattened disk of stars. These galaxies are known as *lenticular galaxies* and are designated *S0*. The *S* notation is perhaps misleading, since these galaxies show no spiral structure. Dwarf S0 galaxies, called *spheroidal galaxies*, are designated *dS0*. They have absolute visual magnitudes fainter than -16 (see Chapter 4). Like dwarf ellipticals, they tend to accompany larger galaxies.

Components of a lenticular galaxy include the *nucleus* (center), *bulge* (surrounding central region), *lens* (or disk; the extension beyond the bulge), and *envelope*, or *halo* (outer extent). These regions do not appear as sharp components in profiles of quantitative light distributions. However, the inner regions are brighter than the outer regions, so the brightness trails off and sharp boundaries are perceived by the eye on an image.

Lenticular designations include subscript numbers following the S0 notation; the numbers refer to the presence of dust or bars. Nonbarred lenticulars may

FIGURE 2.6   *NGC 1201 (top left), NGC 2855 (top right). NGC 2859 (bottom left), NGC 5101 (bottom right). (Images from the STScI Digital Sky Survey.)*

have trace amounts of dust that show up as discrete dust patches. Hubble's numerical designation ranges from SO_1 for nonbarred lenticulars with no dust, through SO_3 for those with a dark band of dust absorption. Barred lenticulars are numbered according to the prominence of a central bar: SBO_1 galaxies have a bar barely protruding from the bulge, and SBO_3 galaxies have narrow and well-defined bars. Examples are shown in Figure 2.6.

Lenticulars, like ellipticals, consist mostly of old low mass stars. These galaxies contain very little gas, so their current rate of star formation is exceedingly small. We will discuss these matters further in Chapters 6 and 10.

2.4 Spiral Galaxies

Bulge/Disk Ratio

Spiral galaxies have bulges in the central regions, thin disks in the outer parts, and a halo surrounding these components. Disks generally are thinner than the lenses of S0 galaxies and show patterns of light concentration called *spiral arms*. Hubble developed a spiral classification system that depended on three parameters: (1) the size of the bulge relative to the disk length, (2) the tightness of the winding of the spiral arms, and (3) the degree of resolution of the disk into stars and H II regions. Often, only the first two properties are considered. The parameters are loosely correlated so that spirals with relatively large central bulges have very tightly wrapped arms and few discernible stars, and spirals with smaller bulges have more-open arms. Spiral galaxies are denoted as *S*, with subdivisions *a*, *b*, or *c*, for galaxies with big bulges and tight windings, down through galaxies with small bulges and more-open arms. The letters *a*, *b*, and so on, are sometimes referred to as the *form family* of the galaxy. The bulge/disk ratio generally is assessed qualitatively through experience. Detailed photometric measurements show that the bulge/disk ratio is not always well-correlated with the tightness of the arms. This sometimes poor correlation can make classification difficult. A quantitative discussion of bulge/disk ratio is in Section 5.4.

The Hubble classification system was expanded and supplemented by Gerard de Vaucouleurs in the 1950s. He added the main divisions Sd and Sm for spiral galaxies, and also added intermediate types Sab, Sbc, Scd, and Sdm. Type Sm was named for "Magellanic spirals" after the prototype Large Magellanic Cloud, which is a dwarf galaxy orbiting our own galaxy and displaying a hint of spiral structure. Figure 2.7 shows examples of the main Hubble-type spiral divisions with de Vaucouleurs' extended divisions.

Sa galaxies have peak disk orbital velocities that are much higher than those of Sc types (averaging nearly 300 km s^{-1} compared with less than 100 km s^{-1} among bright galaxies). The velocity variation with Hubble type leads to many related properties in the disks: The earlier Sa spirals processed their gas more rapidly into stars, and so today are forming fewer stars. Their disks are redder

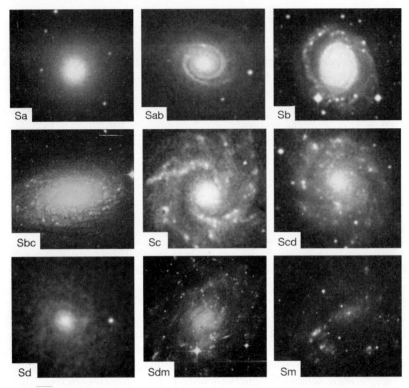

FIGURE 2.7   *The Hubble classification, as extended by de Vaucouleurs, for nonbarred spirals: NGC 7213 (top left), NGC 1357 (top middle), NGC 6753 (top right), NGC 5055 (middle left), NGC 3631 (middle), NGC 3423 (middle right), NGC 4571 (bottom left), NGC 45 (bottom middle), NGC 4395 (bottom right). (Images from the STScI Digital Sky Survey.)*

and more metal-rich than are those of Sc spirals. Rotation and related properties will be considered further in Chapter 7.

Pitch Angle

The tightness of the arm winding is measured by the *pitch angle*, found by drawing a circle at any radius about the nucleus and measuring the angle between the circle and the tangent to a spiral arm (see Figure 2.8). From this definition, we see that tightly coiled arms have very small pitch angles. Pitch angles typically range from about 5° in Sa spirals to about 20° in Sc spirals. In most spirals, the pitch angle is approximately constant for all arms regardless of radius. Variations in pitch angle provide important information about the origins of the spiral structure in a particular galaxy, as will be discussed further in Chapter 5.

Pitch angle

FIGURE 2.8 *Pitch angle is measured at a given radius from a circle to the tangent of the arm.*

Bar Structure

Hubble noted that some spiral galaxies have elongated, bar-shaped stellar configurations. He distinguished these from the "normal" (i.e., nonbarred) spiral galaxies by calling them *SB*, with similar subdivisions as for the normal galaxies. A bar rotates nearly like a solid body, maintaining its shape as the stars orbit the galactic center; the stars move in complicated orbits. The orbits of stars near a bar potential may be elongated either parallel to the bar or perpendicular to it, depending on exactly how far the stars are from the center of the galaxy. We will return to bar properties in Chapters 5 and 9. Examples of barred spiral galaxies for the main Hubble types are shown in Figure 2.9.

Another of de Vaucouleurs' contributions to the Hubble scheme was to recognize intermediate bar types, where the central region is elongated in an oval-type distortion. In this classification, the name SA is given to normal spirals, SB to barred spirals, and SAB to ovally distorted barred spirals. This intermediate classification is an important revision. Formerly, some SAB galaxies were classified as S and some as SB. Dynamically, it turns out that SAB galaxies act more like barred galaxies than like nonbarred galaxies (big bars and ovals are correlated with spiral patterns), so it is often important to differentiate between them and nonbarred galaxies. The *Second Reference Catalogue of Bright Galaxies* (RC2) and the *Third Reference Catalogue of Bright Galaxies* (RC3) list galaxies according to the de Vaucouleurs convention; the *Revised Shapley-Ames Catalog of Bright Galaxies* uses the convention of just S or SB, sometimes using S/SB to designate intermediate types. Examples of SAB galaxies with different Hubble types are shown in Figure 2.10.

Seeking correlations between galaxy properties and Hubble-type parameters is useful in understanding galaxy motions, interactions, and evolution. For graphing or computing, it is often convenient to use numbers rather than letters for different divisions. By convention, astronomers often represent different bar types as 0, 1, 2 for SA, SAB (sometimes written SX), and SB, respectively. The dif-

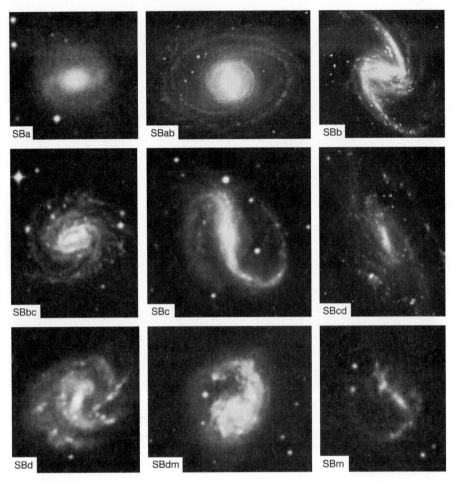

FIGURE 2.9 *Barred spiral (SB) galaxies: NGC 7743 (top left), NGC 1398 (top middle), NGC 1365 (top right), NGC 3660 (middle left), NGC 7479 (middle), IC 5201 (middle right), IC 1953 (bottom left), NGC 4027 (bottom middle), NGC 3664 (bottom right). (Images from the STScI Digital Sky Survey.)*

ferent form families a, ab, b, bc, and so on, may be represented similarly by 1, 2, 3, and up.

Ring Structure

An even more detailed look at spiral structure reveals that arms spiral into the central regions in different ways. Sometimes the arms spiral all the way into the nucleus, and sometimes they spiral in to a ring that circles the nucleus. These

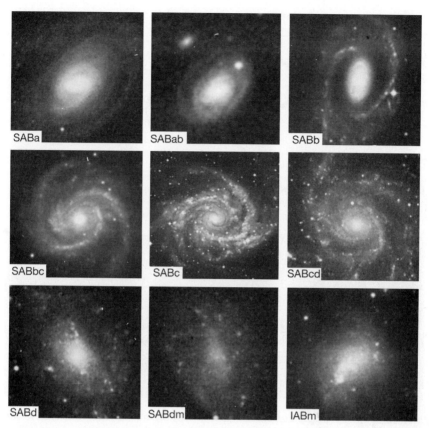

FIGURE 2.10 *Ovally distorted (SAB) galaxies: NGC 1371 (top left), NGC 3368 (top middle), NGC 210 (top right), NGC 4321 (middle left), NGC 2997 (middle), NGC 5457 (middle right), NGC 5585 (bottom left), NGC 4242 (bottom middle), NGC 4214 (bottom right). The designation IABm is discussed in Section 2.6. (Images from the STScI Digital Sky Survey.)*

spiral structures were distinguished by de Vaucouleurs using the notations (s) and (r), respectively. Galaxies with intermediate features are denoted (rs) or (sr). Sketches of different central structures are shown in Figure 2.11, and galaxies are shown in Figure 2.12.

A summary of the de Vaucouleurs subdivisions is shown schematically in Figure 2.13, ranging from nonbarred to barred and from central spiral to central ring structure.

Sometimes galaxies have rings in the outer disks, near the end of the optical spiral arms. These are indicated by an *R* preceding the Hubble type. The RC2 and RC3 include the inner and outer ring designations. Detailed morphologies of different ring structures have been identified by Ron Buta, de Vaucouleurs, and Deborah Crocker. They found that the inner and outer rings may be related to key kine-

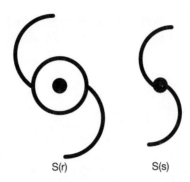

S(r)   S(s)

FIGURE 2.11   *Details of structure in the central regions of spirals.*

Sc(s)   Sbc(rs)   Sc(r)

FIGURE 2.12   *Examples of inner arm and ring structure: NGC 5247 (left), NGC 3344 (middle), NGC 309 (right). (Images from the STScI Digital Sky Survey.)*

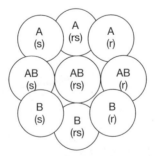

FIGURE 2.13   *Summary of de Vaucouleurs' classification of nonbarred, ovally distorted, and barred spiral types, with central ring or spiral structure.*

matic locations, known as *resonances*, where the disk material rotates synchronously with the spiral pattern. Many galaxies also have nuclear rings, just a few hundred parsecs (pc: 1 pc = 3.1 x 10^{18} cm) from the centers of the galaxies. We will return to a discussion of resonances in Chapter 9, where Figure 9.14 shows an outer ring in NGC 1433 and Figure 9.15 shows a nuclear ring in NGC 3351.

Inclination

Galaxies are randomly inclined to our line of sight, so their apparent shapes vary even if their actual shapes are the same. By convention, the *inclination* of a galaxy is the angle measured between the disk and the perpendicular to our line of sight. Galaxies with inclinations of 90° are called *edge-on*, whereas galaxies with 0° inclination are called *face-on*. Examples are shown in Figure 2.14.

The disks of spiral galaxies are essentially circular when viewed face-on, so an inclined galaxy will have unequal axes. The *Uppsala General Catalogue of Galaxies* (Nilson) lists the apparent axial ratios. The undistorted axis is the major axis, and the shortest, or minor, axis is perpendicular to it. The *Second* and *Third Reference Catalogue of Bright Galaxies*, RC2 and RC3, list the logarithms of the ratios of axes, log R_{25}. (The subscript 25 refers to the light level of 25 mag arcsec^{-2}, which is where the axes are measured; for further description, see Chapter 4.) The inclination, i, can be found from the ratio of axes using geometry:

$$i = \cos^{-1}\left(\frac{1}{10^{\log R_{25}}}\right)$$

or equivalently for major axis a and minor axis b,

$$\boxed{i = \cos^{-1}\left(\frac{b}{a}\right)}$$

FIGURE 2.14 *NGC 3184 (left) has a face-on orientation to our line of sight; NGC 891 (right) has an edge-on orientation. (Images taken with the KPNO 0.6-m Burrell-Schmidt telescope by the author.)*

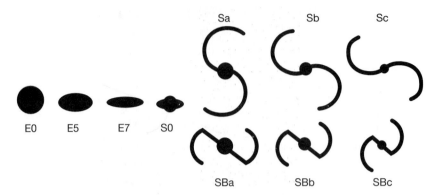

FIGURE 2.15   *Hubble's tuning fork diagram showing the main galaxy types.*

For example, if a major axis is twice as long as the minor axis, then the inclination is $\cos^{-1}(0.5) = 60°$. The logarithm of the axial ratio would then be listed in the RC3 as $\log R_{25} = 0.3$.

2.5 Tuning Fork Diagram

Hubble thought that galaxies might evolve from one form to another, progressing from circular to flattened ellipsoids, to formless lenticular galaxies, to more and more open spirals. Figure 2.15 shows this "tuning fork" diagram.

Several types of observations now indicate that this arrangement is not really an evolutionary sequence. For example, there are equally old stars in all galaxies. Also, ellipticals rotate hardly at all, yet some spirals rotate quickly, and some rotate slowly. The tuning fork diagram remains a useful scheme for organizing the basic types of galaxies, however, and elliptical galaxies are still referred to as *early-type* galaxies and spirals as *early-type spirals* and *late-type spirals* because of this idea. It is theoretically possible, however, for a galaxy to change by half a Hubble type or more, for example from Sab to Sa, through mass redistribution during an encounter with another galaxy. We shall explore these effects in Chapter 12.

Sandage and B. Binggeli have appended a lower-luminosity prong to the fork, paralleling the main structure and ranging from dE to dS0 to Sm and Im, with a possible connection to Blue Compact Dwarfs (see Section 2.6).

2.6 Irregular and Dwarf Galaxies

Irregular galaxies, as their name implies, are galaxies whose shapes do not fit into the previous designations. They tend to be smaller on average than ellipticals or spirals. Irregular galaxies are divided into Type I irregulars (Irr I, now commonly known as Im, or Magellanic irregulars after the Small Magellanic Cloud), and Type II irregulars (Irr II). Type I irregulars are small, slowly rotating systems whose

FIGURE 2.16 *The irregular galaxy NGC 2366. (Imaged in B band with the KPNO 0.9-m telescope by J. Salzer, Wesleyan University and D. Elmegreen.)*

mass is so low that they do not have regular disk structure as do spirals, but they are also flattened systems with a small nucleus. An example is NGC 2366, shown in Figure 2.16. Irregulars may be subdivided into nonbarred (IA), ovally distorted (IAB), or barred (IB) types, just as for spirals.

Many Irr II galaxies are undergoing intense starbursts, presumably as a result of interactions; these effects will be considered in Chapters 10 and 11. M82, the companion to M81, is an example shown in Figure 2.17.

Inclination corrections for irregular galaxies are uncertain because these galaxies are not necessarily circular when viewed from the top. Also, their disks may be thicker than the disks of spirals, so it is not clear which axes to measure for deprojection.

Usually, irregular galaxies are smaller than normal spirals, so they are called *dwarf* galaxies. Other dwarf galaxies do not fit easily into classification schemes. Some dwarf galaxies are extremely blue and appear to be undergoing an intense burst of star formation, although unlike Type II irregulars they do not necessarily have nearby neighbors. These galaxies are called *Blue Compact Dwarfs*, or

FIGURE 2.17 *M82, a companion to M81, is an Irr II galaxy that is undergoing a burst of star formation. Note the streamers perpendicular to the main disk. (Image from the STScI Digital Sky Survey.)*

BCDs; an example is shown in Figure 4.16. The cause of their current level of activity is not well understood but is currently the subject of intense study.

2.7 Peculiar Galaxies

There are some galaxies whose structure or emission is different from normal elliptical, spiral, or lenticular galaxies but is also not irregular in the sense of types Im and Irr II; they are referred to as *peculiar* galaxies. Halton Arp was among the first to identify and catalog them. Sometimes their peculiar morphologies are the result of interactions with neighboring galaxies. Some galaxies are distinguished by their strong radio and nonthermal emission and optical emission lines; these are called *active* galaxies. They will be considered briefly here and in more detail in Chapter 11.

Many active galaxies appear optically as ordinary Hubble types. *Seyferts* are spiral galaxies with strong broad emission from the nucleus, especially from infrared and radio, plus strong variable x-rays. Approximately 1% of all spiral galaxies are classified as Seyferts. They come in two varieties: Type 1 Seyferts have emission lines with broad wings, due to Doppler motions amounting to 1000 to 5000 km s^{-1}, and Type 2 Seyferts have emission lines without broad wings. The active regions in these galaxies produce more infrared than whole (normal) galaxies in the optical. For example, NGC 1068, shown in Figure 2.18, produces an infrared luminosity of 10^{11} L$_o$ (L$_o$ = solar luminosity = 4×10^{33} erg s^{-1}) from a region only 1/1000 the size of the galaxy.

N galaxies have small bright nuclei. They are usually elliptical galaxies. Their classification was not made by Hubble but is based on morphology studied by B. Vorontsov-Vel'yaminov and W. W. Morgan.

Radio lobe galaxies are characterized by the strong radio emission coming from ends of jets 100 kpc (kiloparsec: 1000 pc) to 1 Mpc (megaparsec: 1000 kpc) across. The lobes typically are three or four times the size of the Milky Way. According to theory, they are the result of a strong jet associated with the central galaxy ejection of particles. Examples are shown in Figure 2.19.

FIGURE 2.18   *Seyfert galaxy NGC 1068 looks like an ordinary spiral galaxy. (Image from the STScI Digital Sky Survey.)*

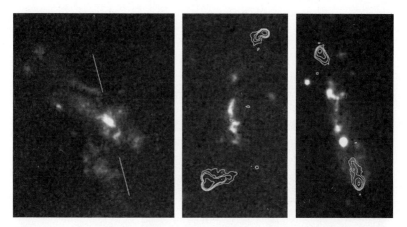

FIGURE 2.19   *Radio lobe galaxies. The contours represent neutral hydrogen (HI) gas detected with the Very Large Array. (left) 3C265 has radio jets along the line, offset from the optical emission; (middle) 3C324 shows several components; (right) 3C368 has knots of star formation or dust. The high-speed gas may be ejected from the vicinity of a central black hole. (Imaged with Hubble Space Telescope WFPC2 by M. Longair, Cambridge; NASA; and NRAO.)*

Quasars are abnormally luminous yet resemble point sources. *Quasar* is an abbreviation for *quasistellar*. There are two types of quasars based on their emission: QSOs have optical emission lines; QSRs have strong radio lines. Recent Hubble Space Telescope observations show that quasars may just be the cores of otherwise ordinary (and sometimes interacting) galaxies, as shown in Figure 2.20. *BL Lac objects*, now popularly known as *blazars*, have no emission lines. They appear to be extreme examples of N-type galaxies and are also stellar in appearance.

All of the above peculiar galaxies have strong nonthermal (i.e., synchrotron) emission, possibly powered by black hole accretion with jets from the disks due to strong magnetic fields. Observations with the Hubble Space Telescope reveal that a large percentage of distant galaxies (which are therefore young) have peculiar and irregular shapes, as shown in Figure 2.21.

There are many other classifications for unusual galaxies. For example, *Zwicky compact objects* are nearly stellar. *Markarian* galaxies have a strong ultraviolet excess. Interacting galaxies are characterized by jets, tails, and rings. *Ocular* galaxies have an inner oval-shaped region and double tidal arms signifying very recent encounters (see Figure 7.6). *Starburst* galaxies have strong ultraviolet and infrared emission in the nucleus, or they may have unusually high star formation in their disks (see Chapter 11). Figure 2.22 shows NGC 1569 in Hα and blue filters; the extra features visible in the Hα image are very young starforming regions. (See Section 3.4 for a discussion of Hα.) Later chapters will illuminate some of the important properties of these galaxies.

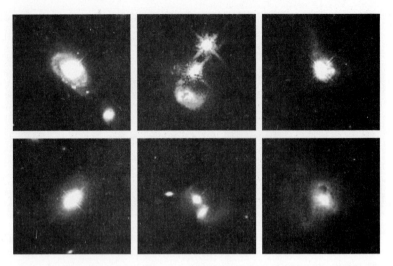

FIGURE 2.20   Quasar host galaxies. (Imaged with Hubble Space Telescope WFPC2 by J. Bahcall, Institute for Advanced Study, Princeton; M. Disney, University of Wales; and NASA.)

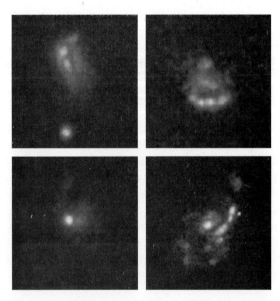

FIGURE 2.21   Irregular and peculiar galaxies. (Viewed with Hubble Space Telescope WFPC2 by R. Griffiths, John Hopkins University; The Medium Deep Survey Team; and NASA.)

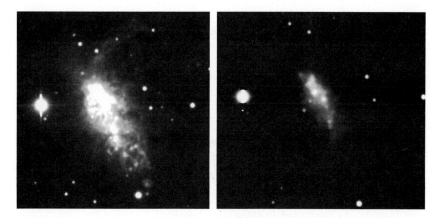

FIGURE 2.22 *NGC 1569 in Hα (left) and blue light (right). (Images from the KPNO 0.9-m telescope, by J. Salzer, Wesleyan University and D. Elmegreen.)*

2.8 Catalogs and Atlases

Atlases and catalogs of galaxies provide compilations of properties that are useful for galaxy research. The position of a galaxy in the sky generally is designated by one of two coordinate systems. By convention, optical astronomers generally use the *Equatorial Coordinate System*, whose coordinates are based on the rotation of Earth. The *first point of Aries*, where the ecliptic crosses the celestial equator (which is an extension of Earth's equator), is the zero point for *right ascension* (RA, or α), which increases from west to east in the sky and is like an extension of the longitude system used on Earth. Right ascension is measured in hours, minutes, and seconds (h, m, s). *Declination* (Dec, or δ), is the elevation of a source above (+) or below (−) the celestial equator; it is measured in degrees, arcminutes, and arcseconds (°, ', "). Declination is measured from 0° to 90° north or south of the equator, so it is an extension of latitude onto the celestial sphere. This coordinate system is sketched in Figure 2.23. Because of Earth's precession, the *epoch* must be specified; generally, the reference system is based on A.D. 1950 coordinates or A.D. 2000 coordinates.

Radio astronomers, by convention, use the *Galactic Coordinate System*. In this system, the plane of our Galaxy, rather than Earth's equator, is the reference point for *galactic latitude*, abbreviated as *b* and measured in degrees. The *galactic longitude, l*, is measured in degrees counterclockwise from the center of the galaxy. This coordinate system has been revised once, so most coordinates are specified in the revised system as l^{II}, b^{II}. This coordinate system is sketched in Figure 2.24.

There are many numbering conventions for galaxies. Most commonly, bright galaxies are referred to by their NGC, or *New General Catalogue*, number. This

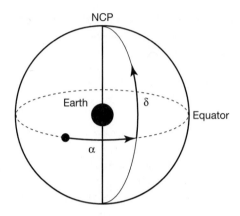

FIGURE 2.23   *Equatorial coordinate system. Declination is measured from the celestial equator, which is an extension of Earth's equator. NCP is the North Celestial Pole, an imaginary extension of Earth's North Pole. Right ascension is measured counterclockwise from the first point of Aries.*

number stems from a listing of 5079 nebulous objects by John Herschel in 1864 and was revised by J. L. E. Dreyer in 1888 to include 7840 objects. Charles Messier's list of nebulous objects includes some of the NGC objects, so these have NGC numbers as well as M (Messier) numbers. *The Revised New General Catalogue of Nonstellar Astronomical Objects* by J. Sulentic and W. Tifft provides descriptions of each NGC object. In addition, galaxies might be listed by their IC (*Index Catalogue*) numbers, which were two separate sets of additions that Dreyer made to his catalog in 1895 and 1898, or by their Zwicky number or their MCG (*Morphological Catalog of Galaxies*) or UGC (*Uppsala General Catalogue of Galaxies*) numbers. More obscure galaxies are listed by their anonymous designation, *Ahhmmdd*, in which the *hhmm* indicates the right ascension in hours and minutes and *dd* indicates declination in degrees (see below).

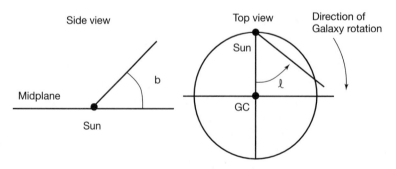

FIGURE 2.24   *Galactic coordinate system. GC, Galactic Center.*

The *Morphological Catalog of Galaxies* (MCG) was compiled by Vorontsov-Vel'yaminov and collaborators over a 12-year period. It contains a list of 32,000 galaxies based on the *Palomar Observatory Sky Survey* (POSS). The POSS was first made with the 1.2-m Schmidt telescope at Palomar Mountain in the early 1950s. It consists of approximately 2000 plates, each about 6° x 6°, taken of the entire Northern Hemisphere. The survey was made in the blue (designated as O plates because the emulsions are called 103a-O emulsions) and in the red (designated as E plates because the emulsions are called 103a-E). The survey was re-done in B, R, and I passbands (see Section 3.6) in the 1980s and 1990s, partly to assist in defining star coordinates for the Hubble Space Telescope guidance system. The European Southern Observatory (ESO) has also completed a Southern Hemisphere survey in B and R bands. These surveys, formerly available as glass plates or prints, are now being digitized and produced as CD-ROMs for convenience in research. The Space Telescope Science Institute (STScI) has made the Digital Sky Survey available through the World Wide Web (see Further Reading at the end of the chapter).

There are many galaxy atlases that are useful for studying detailed morphology; their complete references are listed in Further Reading at the end of the chapter. Among the more notable are several by Sandage and collaborators: the *Hubble Atlas of Galaxies*, the *Revised Shapley-Ames Catalog of Bright Galaxies* (RSA), the *Carnegie Atlas of Galaxies* (Sandage and Bedke), the *Atlas of Galaxies Useful for Measuring the Cosmological Distance Scale*. Other important atlases include H. Arp's *Atlas of Peculiar Galaxies,* and J. Wray's *The Color Atlas of Galaxies*. The latter is particularly useful in showing the true colors of galaxies by combining images in several filters. There are also smaller atlases of galaxies in the literature, such as the Sandage and B. Brucato "Atlas of Southern Galaxies," the "Near-Infrared Atlas of Galaxies" by D. Elmegreen, the "Catalogue of Southern Ring Galaxies" by Buta, and the "Morphology of Low Surface Brightness Disk Galaxies" by J. McGaugh, R. Schombert, and G. Bothun.

R. B. Tully has compiled a list of galaxy parameters, most notably a standardized system of recession velocities and distances, in his *Nearby Galaxies Catalog* and *Nearby Galaxies Atlas*. The atlas consists of color-coded velocity information so that neighboring galaxies can be readily identified. F. Zwicky has compiled a six-volume catalog of groups and clusters of galaxies. Several papers in the literature have also sought to identify groups and clusters of galaxies by various objective parameters; these include the J. Huchra and M. Geller list, binary pairs listed by S. Peterson, and small groups by P. Hickson. The *Catalogue of Radial Velocities of Galaxies* by G. Palumbo, G. Tanzella-Nitti, and G. Vettolani is useful for identifying neighbors also.

Basic galaxy parameters such as coordinates, size, shape, brightness, color, velocity of recession, inclination, and orientation are included in the RSA and in the reference catalogs by de Vaucouleurs and colleagues (*Second Reference Catalogue of Bright Galaxies* [RC2] and *Third Reference Catalogue of Bright Galaxies* [RC3]). We will return to many of these properties as we study them in detail in subsequent chapters. The *Uppsala General Catalogue of Galaxies* (UGC) by P. Nilson also has similar useful data.

Exercises

1. Read pages 1–25 of Sandage's *Hubble Atlas of Galaxies* and pages 1–58 of Sandage and Bedke's *Carnegie Atlas of Galaxies*, vol. 1, for a history of Hubble's classification system and the underlying motivations.

2. Look up the following galaxies in one of the standard catalogs (such as RC3) or on computer (see Useful Websites) to determine their type and coordinates: NGC 157, NGC 5248, NGC 7479.

3. Locate the galaxies in Exercise 2 on the POSS (Palomar Observatory Sky Survey): Note that N is up, E is left since we're looking at the sky as if from inside a fishbowl. Become familiar with degrees, minutes, and seconds by using the transparency overlays to get a sense of the scale. Note the difference in appearance of galaxies on the O (blue-sensitive emulsion) and E (red-sensitive emulsion) prints. These galaxies can also be imported to your computer via the websites listed below.

4. Find the galaxies in Exercise 2 on the Tully maps or POSS prints and see if they have neighbors within 10 galaxy diameters.

5. Look up the following nearby galaxies in any of the atlases; you should learn to identify them and know their approximate Hubble types. (Most of the exercise may also be done by importing the images via the websites.) Fill in the table with the following information in each column.

   a. NGC number and alternative designation, if any.

   b. Atlas name and page number of photo.

   c. Hubble classification from the *Third Reference Catalogue of Bright Galaxies* (RC3) *and* from Sandage and Tammann's *Revised Shapley-Ames Catalog of Bright Galaxies*; they are probably different.

   d. Measure the bulge and disk to determine the bulge/disk ratio.

   e. Measure the approximate pitch angle of the arms (do this by drawing a circle at any radius. Draw a tangent to the spiral arm and use a protractor to measure the angle between the tangent and the circle; you may want to measure it at a few radii and take an average). Note that if the galaxy is not face-on, you need to measure close to the major axis.

   f. Determine the inclination in degrees by measuring the major and minor axes; record both of these and the inclination you derive.

   g. Determine the inclination in degrees using, e.g., RC2 col. 11 or RC3, making use of the convention for inclination as follows:

$$i = \cos^{-1}\left(\frac{1}{10^{\log R_{25}}}\right)$$

For example, if $\log R_{25} = 0.3$, $i = 59.9°$.

NGC and other designation	Atlas, page	Hubble type RC3	RSA	Bulge Disk Ratio	Pitch angle	Major axis Minor axis Inclin.	log R_{25} Inclin.
628							
925							
1232							

NGC and other designation	Atlas, page	Hubble type RC3 RSA	Bulge Disk Ratio	Pitch angle	Major axis Minor axis Inclin.	log R_{25} Inclin.
1300						
2841						
3031						
4321						
5055						
5194						
5457						

Unsolved Problems

1. Can computers be used to provide accurate classifications of galaxies?

2. What are the physical variables that lead to different galaxy morphologies?

Useful Websites

The Space Telescope Science Institute (STScI) has press releases and public photos available through http://www.stsci.edu.

The Space Telescope Science Institute has also made the Digital Sky Survey accessible through http://archive.stsci.edu/dss to download or view the object of your choice to your computer. Coordinates in different epochs can also be obtained.

A useful site for general literature searches is http://adsabs.harvard.edu/abstract_service.html.

A good site for galaxy information is the National Extragalactic Database (NED). Access it by telnet to ned.ipac.caltech.edu, login ned, or on the Web by http://ned.ipac.caltech.edu.

A catalog of 113 galaxy images in blue and red obtained with the Palomar 1.5-m and Lowell 1.1-m telescopes is available through http://astro.princeton.edu/~frei/galaxy-catalog.html.

Further Reading

Arp, H. C. 1966. *Atlas of peculiar galaxies.* Pasadena: California Institute of Technology.

Arp, H. C., and B. Madore. 1987. *Catalogue of southern peculiar galaxies and associations.* Cambridge: Cambridge University Press.

Buta, R. 1995. The catalogue of southern ring galaxies. *Astrophysical Journal (Suppl.)* 96:39.

Buta, R., and D. Crocker. 1992. Integrated photometric properties of early type ringed galaxies. *Astronomical Journal* 103:1804.

Buta, R. and D. Crocker. 1993. Metric characteristics of nuclear rings and related features in spiral galaxies. *Astronomical Journal* 105:1344.

de Vaucouleurs, G. & R. Buta. 1980. Diameters of nuclei, lenses, and inner and outer rings in 532 galaxies. *Astronomical Journal* 85:637.

de Vaucouleurs, G., A. de Vaucouleurs, and H.G. Corwin. 1976. *Second reference catalogue of bright galaxies.* Austin: University of Texas Press.

de Vaucouleurs, G., A. de Vaucouleurs, H.G. Corwin, R. Buta, G. Paturel, and P. Fouque. 1991. *Third reference catalogue of bright galaxies,* New York: Springer.

Elmegreen, D.M. 1981. Near-infrared atlas of galaxies. *Astrophysical Journal (Suppl.)* 47:229.

Gallagher, J., and D. Hunter. 1984. Structure and evolution of irregular galaxies. *Annual Review of Astronomy and Astrophysics* 22:37.

Geller, M. and J. Huchra. 1983. Groups of galaxies. III. The CfA survey. *Astrophysical Journal (Suppl.)* 52:61.

Hickson, P., E. Kindl, and J. Auman. 1989. A photometric catalog of compact groups of galaxies. *Astrophysical Journal (Suppl.)* 70:687.

Huchra, J., and M. Geller. 1982. Groups of galaxies. I. Nearby groups. *Astrophysical Journal.* 257:423.

Kormendy, J. 1978. A morphological survey of bar, lens, and ring components in galaxies. Secular evolution in galaxy structure. *Astrophysical Journal* 227:714.

McGaugh, S., J. Schombert, and G. Bothun. 1995. The morphology of low surface brightness disk galaxies. *Astronomical Journal* 109:2019.

Nilson, P. 1973. *Uppsala general catalogue of galaxies.* Uppsala: Royal Society of Sciences of Uppsala.

Palumbo, G., G. Tanzella-Nitti, and G. Vettolani. 1983. *Catalogue of radial velocities of galaxies.* New York: Gordon and Breach Science Publishers.

Peterson, S. 1979. Double galaxies. I. Observational data on a well-defined sample. *Astrophysical Journal Supplement Series* 40:572.

Roberts, M.S., and M.P. Haynes. 1994. Physical parameters along the Hubble sequence. *Annual Review of Astronomy and Astrophysics* 32:115.

Sandage, A. 1961. *The Hubble atlas of galaxies.* Washington, D.C.: Carnegie Institution of Washington.

Sandage, A., and J. Bedke. 1988. *Atlas of galaxies useful for measuring the cosmological distance scale.* Washington, D.C.: NASA Scientific and Technical Information Division.

Sandage, A., and J. Bedke. 1994. *The Carnegie atlas of galaxies.* Vols. 1 and 2. Washington, D.C.: Carnegie Institution of Washington and The Flintlock Foundation.

Sandage, A., and B. Binggeli. 1984. Studies of the Virgo cluster. III. A classification system and an illustrated atlas of Virgo cluster dwarf galaxies. *Astronomical Journal* 89:919.

Sandage, A., and R. Brucato. 1979. Atlas of southern galaxies. *Astronomical Journal* 84:472.

Sandage, A., and G. Tammann. 1981. *A revised Shapley-Ames catalog of bright galaxies.* Washington, D.C.: Carnegie Institution of Washington.

Sulentic, J., and W. Tifft. 1973. *The revised new general catalogue of nonstellar astronomical objects*. Tucson: University of Arizona Press.

Tully, R.B. 1988. *Nearby galaxies atlas*. Cambridge: Cambridge University Press.

Tully, R.B. 1988. *Nearby galaxies catalog*. Cambridge: Cambridge University Press.

Turner, E., and J.R. Gott. 1976. Groups of galaxies. I. A catalogue. *Astrophysical Journal (Suppl.)* 32:409.

van den Bergh, S. 1990. S0 galaxies and the Hubble sequence. In *Astronomical Society of the Pacific Conference Series 10: Evolution of the universe*, San Francisco: Astronomical Society of the Pacific, ed. R. Kron, 70.

van den Bergh, S., M. Pierce, and R. Tully. 1990. Classification of galaxies on CCD frames. *Astrophysical Journal*. 359:4.

Vorontsov-Vel'yaminov, B.A., V. Arkhipova, and A. Krasnogorskaya. Part 1, 1962; Part 2, 1964; Part 3, 1963; Part 4, 1968; Part 5, 1974. *Morphological catalog of galaxies*. Moscow: Sternberg Observatory.

Wray, J.D. 1988. *The color atlas of galaxies*. Cambridge: Cambridge University Press.

Zwicky, F., E. Herzog, and P. Wild. 1961–1968. *Catalogue of galaxies and clusters of galaxies*, vols. 1–6. Pasadena: California Institute of Technology.

CHAPTER 3
Star Light

Chapter Objectives: to describe stellar line and continuum radiation and measurements

---

Toolbox:

Planck function

intensity

Wien's Law

flux

Stefan's law

magnitude

distance modulus

spectral transitions

Boltzmann equation

Saha equation

filters

colors

3.1 Blackbody Radiation and the Electromagnetic Spectrum

Galaxies emit radiation all along the electromagnetic spectrum, from high frequencies to low frequencies. We will review radiation properties in order to interpret the observed spectral distributions of different galaxies. Most of the radiation from normal galaxies such as the Milky Way comes from stars, which radiate in the optical part of the spectrum, where our eyes are most sensitive, so their radiation will be examined first.

Stars radiate approximately like *blackbodies*, which are hypothetical perfectly absorbing and perfectly radiating bodies. Their radiation is represented by the formula known as the *Planck function* or *Planck law*:

$$I = \frac{2h\nu^3}{c^2} \frac{1}{e^{h\nu/kT} - 1} \text{ erg cm}^{-2} \text{ s}^{-1} \text{ Hz}^{-1} \text{ ster}^{-1}$$

for intensity I, frequency ν, temperature, T, and constants h, c, and k (Appendix 1).

The units of intensity indicate that it is an energy (ergs) per unit time (s^{-1}), or a power, distributed over an area (cm^{-2}) throughout a given *solid angle*

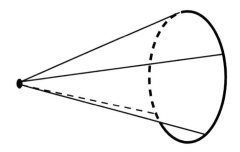

FIGURE 3.1   *A solid angle is a three-dimensional angle.*

(ster, or *steradian*) in a given frequency interval v to $v + \Delta v$ (Hz = Hertz = cycles per second). A solid angle is a three-dimensional angle, as shown in Figure 3.1. A steradian is the measure of a solid angle. Just as there are 2π radians in a circle, and a radian is equal to 57.3°, so there are 4π steradians in a sphere, and a steradian is $57.3^2 = 3283$ square degrees.

A plot of blackbody intensity as a function of wavelength follows a characteristic shape known as the *Planck curve*, as shown in Figure 3.2.

As is evident from Planck's law and Figure 3.2, all blackbodies radiate some at all wavelengths. Furthermore, a hotter blackbody radiates more *at all wavelengths* than a cooler blackbody. This property means that a hot star that appears blue is also radiating more in the red part of the spectrum than is a cool star that appears red.

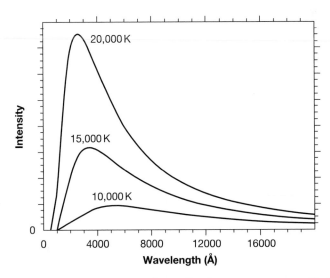

FIGURE 3.2   *Blackbody curves are shown for three temperatures (in kelvins).*

By differentiating Planck's law, we find that the wavelength of the most intense blackbody emission is inversely proportional to the temperature of the radiating body. This relation is apparent from the three curves in Figure 3.2, where the 20,000 K blackbody peaks at shorter wavelengths than the cooler blackbodies. This relation is most conveniently remembered in the form of *Wien's law*:

$$\lambda_{peak} = \frac{C}{T}$$

where λ_{peak} is the peak emission wavelength for a blackbody radiating at a temperature T. The constant of proportionality, C, has a value of 2.9×10^7 K-Å when the wavelength is measured in angstroms (Å) and the temperature is in kelvins. Figure 3.3 shows Wien's law for some prominent stars: Sirius and Rigel, very hot stars, are noticeably blue-white, whereas Betelgeuse, a cooler star, is red. Our Sun is yellow because its surface temperature is 5850 K.

We derive the total energy flux, F, by integrating Planck's function for intensity over solid angle and frequency. This relation is known as *Stefan's law*:

$$F = \sigma T^4 \text{ erg cm}^{-2} \text{s}^{-1}$$

where T is the surface temperature of the star measured in kelvins (K), and σ is the Stefan-Boltzmann constant, equal to 5.67×10^{-5} erg K^{-4} cm^{-2} s^{-1}. Sometimes flux is referred to as "brightness." As is evident from the units, the flux measures the amount of energy passing through a surface in a given time.

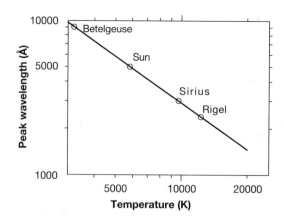

FIGURE 3.3   *Wien's law shows that cooler stars are redder than hotter stars.*

A star's luminosity, L, is the product of its surface area and flux:

$$L = 4\pi R^2 \sigma T^4 \text{ erg s}^{-1}$$

where R is the radius of the star.

As an example of the previous units, consider a star with a radius of 7×10^{10} cm that has a surface temperature of 6000 K (similar to the Sun). Its flux is

$$F = (5.67 \times 10^{-5})6000^4 = 7.35 \times 10^{10} \text{ erg cm}^{-2}\text{ s}^{-1}$$

Its luminosity is

$$L = 4\pi R^2 F = 4\pi(7 \times 10^{10})^2(7.35 \times 10^{10})$$
$$= 4.5 \times 10^{33} \text{ erg s}^{-1}$$

and its intensity at 5000 Å is

$$I = \frac{2 \times (6.63 \times 10^{-27}) \times (3 \times 10^{10})}{(5000 \times 10^{-8})^3 \times (e^{h\nu/k6000} - 1)}$$
$$= 2.63 \times 10^{-5} \text{ erg cm}^{-2}\text{ s}^{-1}\text{ Hz}^{-1}\text{ ster}^{-1}$$

3.2 Magnitudes

In order to quantify a star's brightness, we need to establish a measurement scale. The eye has an approximately logarithmic response to light, meaning that it perceives brightness *ratios* as *differences* in brightness. The *magnitude* scale, developed by Hipparchus in the second century A.D. for naked-eye observations, makes use of this fact. By convention, magnitude, m, and flux, F, are defined as follows:

$$m_\lambda = -2.5 \log F_\lambda + \text{constant}$$

where the subscript specifies a particular wavelength or passband. (The constant contains an arbitrary zero-point.) Then for two wavelengths 1 and 2 (or two stars 1 and 2), we measure a magnitude difference

$$m_1 - m_2 = -2.5 \log \frac{F_1}{F_2}$$

The magnitude scale thus has equal magnitude differences that are related to equal flux ratios. From the definition, we can see that a difference of 5 magnitudes is exactly equal to a factor of 100 in brightness. That is, if $m_1 - m_2 = 5$, then $F_2/F_1 = 100$. For any magnitude difference Δm, the flux ratio as shown in Table 3.1, is given approximately by $(2.512)^{\Delta m}$, and exactly by $10^{0.4\Delta m}$.

The brightness of an object as it appears to us on Earth is known as its *apparent magnitude*. Stars seen with the unaided eye have apparent magnitudes between -1 and 6 mag. Note from the definition that brighter objects have smaller magnitudes, so a first-magnitude star is 100 times as bright as a sixth-magnitude star. Bright nearby galaxies beyond our Local Group (see Chapter 12) have integrated apparent magnitudes of about 10 mag, which is about 40 times too faint to see by eye.

TABLE 3.1   *Brightness comparisons*

Magnitude difference	Flux ratio
1	2.512
2	6.310
3	15.85
4	39.81
5	100
10	10^4
15	10^6
20	10^8

3.3 Distance Modulus

The apparent brightness of an object diminishes as the inverse square of our distance from it: $F_A/F_B \sim (r_B/r_A)^2$. This relation makes sense when the units of flux are considered: Flux depends on the surface area over which an object's radiation is spread, and the surface area increases as the square of the distance from the object. Thus, for two identical objects A and B at arbitrary distances r and $2r$, respectively, object A will have a flux four times as bright as object B. This relation is evident in Figure 3.4. Imagine that the dot is a star. The solid angle has an amount of light L_A passing through it at the first square. The four squares a distance $2r$ away are all the same size as the first square, but only one-fourth L_A passes through each of those squares (so that the total luminosity, $4 \times 1/4 L_A = L_A$, still passes through the solid angle). Thus, the flux through each square has decreased as the inverse square of the distance.

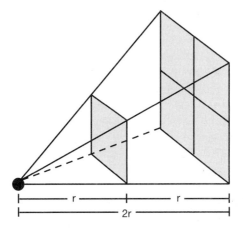

FIGURE 3.4 *All squares are the same linear size, but each square at 2r receives only one-fourth the light going through the first square.*

The magnitude difference can be rewritten in terms of distance:

$$m_1 - m_2 = 2.5 \log\left(\frac{r_1}{r_2}\right)^2 = 5 \log\left(\frac{r_1}{r_2}\right)$$

An object's intrinsic brightness is known as its *absolute magnitude*. By convention, the reference distance that defines the absolute magnitude is 10 pc. The parsec derives its name from the way in which it is measured. The apparent displacement of a nearby star relative to the background stars as Earth orbits the Sun is known as a *parallax*. If the displacement of the star is observed over a six-month interval, when Earth is on opposite sides of the Sun, then the corresponding distance to the star can be determined from trigonometry. The Earth-Sun baseline is equal to 1 astronomical unit (AU), or 1.5×10^{13} cm, and the parallactic angle, P, is indicated in Figure 3.5.

Figure 3.6 indicates how parallactic motion would show up in an astronomical image. The small dots represent some distant background stars that appear to be fixed in the sky relative to one another. A single nearby star appears to shift its position with respect to them as Earth travels around the Sun. The displacement from one time to the next is measured as an angle in the sky.

If the parallax amounts to 1 arcsecond, the distance between Earth and the star is called a parsec, an abbreviation for parallax of 1 second of arc. The kiloparsec, and the Megaparsec (1000 kpc), Mpc, are particularly appropriate for describing the sizes of galaxy disks and the distances to galaxies, respectively. The method of trigonometric parallax is useful only for nearby stars less than about 100 pc away because of limitations due to seeing and resolution. In general, we have a distance d in parsecs given by the inverse of the parallactic angle measured in arcseconds (for a baseline of 1 AU; " is the notation for arcsecond):

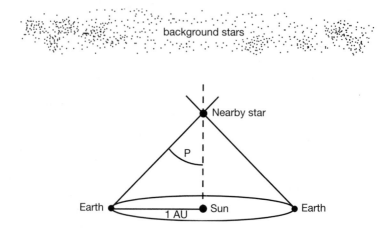

FIGURE 3.5 *Parallax angle P is given by the shift of a nearby star with respect to very distant background stars as Earth orbits the Sun.*

$$d(pc) = \frac{1(AU)}{P('')}$$

Then an object's apparent magnitude, m, compared with its absolute magnitude, M, depends on its distance from us compared with the reference distance of 10 pc:

$$m - M = 5 \log\left[\frac{r\,(pc)}{10(pc)}\right]$$
$$= 5 \log r\,(pc) - 5$$

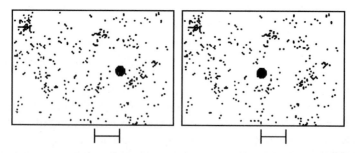

FIGURE 3.6 *A nearby star's parallax shows up as a shift (greatly exaggerated here) on an image taken at two different times of the year. The parallax angle P is half of the total shift (as in Figure 3.5). The nearby star is indicated by a large dot; background stars are indicated by small dots.*

This equation, known as the *distance modulus*, is one of the most important equations in galactic astronomy. If the absolute magnitude of an object is known, then a measure of its apparent magnitude allows its distance to be determined. Details including complications due to the presence of dust will be explored further in Chapter 4.

3.4 Stellar Spectra

Although stars behave like blackbodies to first approximation, their spectra are replete with emission and absorption lines. Here we review briefly the physical processes of line formation. Different types of atoms have different numbers of electrons. The electrons are configured in orbits whose radii are quantized; that is, only certain average distances from the nucleus and certain orbit shapes are allowed. These distances and shapes are specified by the laws of quantum mechanics and depend on the spin and orbital angular momentum of the electrons. Transitions are dictated by the wavelike properties of an electron. An *allowed transition* is one in which an integral number of wavelengths can be completed in one electron orbit. In the simple Bohr model for hydrogen, in which the proton is orbited by an electron a given distance away, the angular momentum, L, is given by the equation:

$$L = mvr = \frac{nh}{2\pi}$$

which specifies the allowed distances r for energy levels n; m is the mass of the atom, and h is Planck's constant (see Appendix 1).

Electrons can move to different average distances from the nucleus when energy is exchanged. Collisions between the atom and another particle, or absorption of an incoming photon by the atom, can transfer energy and cause an electron to jump from one energy level to another. A jump to a higher average radius is known as an *excitation*. The excited electron tries to reestablish its lowest energy state, which it can do through another collisional transfer of energy or through the release of a photon of frequency ν whose energy corresponds to the energy difference $\Delta E = E_n - E_m = h\nu$ between the excited level (n) and de-excited level (m). When the photon is released, it contributes to the emission line spectrum that is characteristic of any given atom. The inverse process, in which an electron jumps to a higher energy level, requires the absorption of energy, as shown in Figure 3.7.

Hydrogen is both the simplest and the most abundant atom in the Universe. Most stars are about 90% hydrogen by number or about 74% hydrogen by mass. The spectrum of hydrogen includes many different lines corresponding to all of the possible jumps that an electron can make. The most commonly observed line involves a jump from the third to the second energy level, which happens

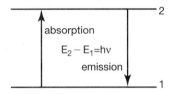

FIGURE 3.7   *Absorption and emission between two levels.*

to have an energy difference corresponding to a red photon, with a wavelength of 6563 Å. Most of the red color in astronomical photographs of nebulae results from this emission line. When an electron jumps one energy level, the line is referred to as the α line; a jump of two energy levels is a β transition, and so on; the 6563 Å line is represented as Hα. The series of hydrogen lines that arise in the visible part of the electromagnetic spectrum is known as the *Balmer series* and involves jumps into and out of the second energy level. Jumps into and out of the first level require higher energies; the resulting ultraviolet lines are known as the *Lyman* series. Jumps into and out of the third level give rise to the infrared *Paschen* series.

The wavelength of any transition in hydrogen is specified by the relation

$$\frac{1}{\lambda} = R\left(\frac{1}{m^2} - \frac{1}{n^2}\right) \text{cm}^{-1}$$

where R is the Rydberg constant, 1.097373×10^5 cm^{-1}, m is the lower level (in this case 2), and n is the upper level. The value of R was first determined empirically by J. Balmer and later became understood from basic principles of quantum mechanics. Figure 3.8 shows the lines in the Balmer series.

In an ensemble of atoms, such as the atmosphere of a star, the strength of a given emission line depends on the number of electrons in the upper and lower

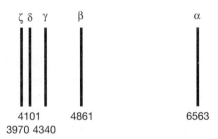

FIGURE 3.8   *The Balmer series of hydrogen is shown; wavelengths are given in angstroms.*

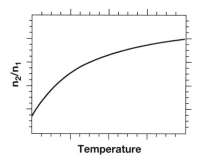

FIGURE 3.9 *The Boltzmann equation gives the ratio of populations of two states; it approaches the ratio of the statistical weights at high temperatures.*

levels and the temperature of the gas, which are related through the probabilistic formula known as the *Boltzmann equation*:

$$\frac{n_2}{n_1} = \frac{g_2}{g_1} e^{-h\nu/kT}$$

Here n_1 and n_2 represent the populations of the lower and upper levels of an electron transition; g_1 and g_2 are called *statistical weights*, ν is the frequency of the transition (so that the energy difference of two levels, E, is given by $E = h\nu$), and T is the thermal energy associated with the transition. Statistical weights are determined by the laws of quantum mechanics. They specify the number of subdivisions of an energy level, given by $g = 2J+1$, where J is the quantum number for the total electron angular momentum. In the presence of a magnetic field, the lines in an energy level would split; these are known as *fine structure lines*. They have been measured for different transitions in different elements and are listed in various references (see Allen 1973). Figure 3.9 shows how the ratios of populations of two levels approach a constant value (equal to the ratio of the statistical weights) at high temperatures because of the exponential term.

3.5 Stellar Classification

Stars commonly show Hα and Hβ absorption lines, because their cool outer photospheric layers absorb photons from the underlying layers. Some stars have very strong Balmer lines, and others do not. Annie Jump Cannon, a Wellesley graduate who worked at Harvard in the early 1900s, originally classified stars according to the strength of these lines because they are so prominent in many stars. The stars with the strongest lines were given the designation A, slightly weaker were called B stars, and so on. Other lines were used for the classifica-

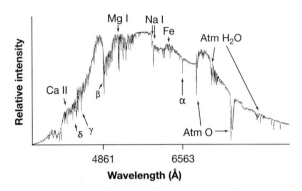

FIGURE 3.10 *Solar spectrum recorded with an Oriel spectrograph. Some lines are marked, including the hydrogen Balmer lines and lines caused by Earth's atmosphere. (By D. Kominsky, L. Ruocco, and D. Hasselbacher, Vassar College.)*

tion, too; K and M stars, for example, had prominent metallic lines. Figure 3.10 shows a low-resolution spectrum of the Sun, a G star.

It was not until M. Saha in 1921 developed an equation relating ionizations to temperatures that astronomers realized that the absorption lines seen in stellar spectra were a result of different surface temperatures. The *Saha equation* for ionization is:

$$\frac{n_e n(X_{r+1})}{n(X_r)} = \frac{2g_{r+1}}{g_r}\left(\frac{2\pi m_r k\, T_k}{h^2}\right)^{3/2} \exp\left[\frac{-E_i}{k\, T_k}\right]$$

These terms are similar to the Boltzmann equation, except for the additional term n_e for the electron density of the gas, and the $T^{3/2}$ term, which results from electron degrees of freedom. The subscripts of X refer to the r+1 and r stages of ionization (that is, the number of electrons that have been removed from the atom). E_i is the energy required to go from one ionization state to the next. (For hydrogen, there is only a single electron, so only one ionization is possible.) Ionization of a hydrogen atom from a level n requires an energy of $E_n = -13.6$ eV/n^2, where eV is an electron volt, equal to 1.602×10^{-12} erg.

With this understanding of the cause of spectral lines, astronomers rearranged the order of the different stellar spectra to represent a temperature sequence rather than an absorption line sequence, because temperature ordering has a physical significance. Nearly 200,000 stars had been classified at that point, so it was easier to rearrange the alphabetical sequence than to relabel all the stars. Thus, we now know the stellar sequence as O, B, A, F, G, K, M. The mnemonic device normally associated with this sequence, for better or for worse, is "Oh, Be A Fine Girl (or Guy), Kiss Me." Some astronomers originally and incorrectly thought that the temperature sequence represented an evolutionary sequence, so that O and B stars are still referred to as "early type," and G, K, and M are referred to as "late type," just as

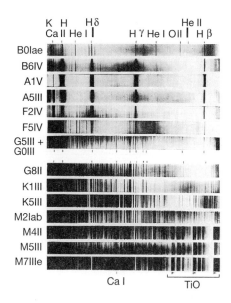

FIGURE 3.11 *Stellar spectra showing the main types. (From the University of Michigan.)*

galaxies are. The spectral types are subdivided into 10 further groups; there are B0–B9 stars, followed by A0, A1, . . . A9, and so on. Figure 3.11 shows spectra of the main stellar types. Note how much more complicated the spectra of the later type stars are, because so many heavy atoms are excited at low temperatures.

Stars with the strongest Hα lines are not the hottest stars. The A0 stars have the strongest Hα lines, because their 10,000 K surface temperatures are just right to maximize the 2–3 transition for Hα absorption. O stars have hot surface temperatures as high as 30,000 K to 50,000 K, so the hydrogen (as well as helium) is ionized. This high temperature means that there are few 2–3 transitions. The cooler stars, such as our Sun, a G star with a surface temperature of 6000 K, are not hot enough to pump many hydrogen electrons up to a higher level, so again there are few 2–3 transitions in these stars. Instead, they show strong lines of calcium and iron and other ionized metals. The coolest M stars, with surface temperatures of 3000 K, have strong neutral metals as well as some molecular lines from (TiO) and (CN) in the outer stellar atmospheres. Prominent ions for different spectral types are shown in Figure 3.12.

3.6 Filters

Broadband filters are useful for measuring radiation from stars, because they quickly sample chunks of the blackbody curves; note that two points on a Planck curve are sufficient to define the star's temperature precisely in the absence of dust. The process of determining the magnitude as a function of filter is known

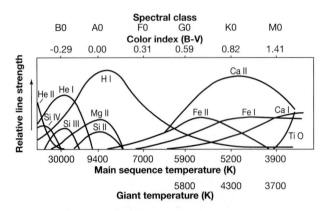

FIGURE 3.12 *The relative strengths of different ions in stellar spectra depend on the stars' surface temperatures. Some of the prominent lines are shown here. (Adapted from Kaler 1989.)*

as *photometry*. There are several photometric systems of filters in standard use. In general, the filters are a few hundred angstroms in width and peak approximately at 3300 Å in ultraviolet (U), ~ 4400 Å in blue (B), ~5500 Å in visual (V), ~6400 Å in red (R), and ~8800 Å in infrared (I), with variations depending on the filter system. Table 3.2 lists some commonly used filters.

TABLE 3.2 *Photoelectric filter system*

	U	V	B	G	R	I	
Six-color Stebbins Whitford Kron	3550	4200	4900	5700	7200	10300	
	U	B	V				
UBV (Johnson Morgan)	3650	4400	5500				
	U	B	V	R	I		
Cousins	3960	4330	5470	6400	7900		
	R	I	J	K	L	M	N
Infrared (Johnson)	7000	8800	1.25μ	2.2μ	3.4μ	5.0μ	10.4μ
	u	v	b	y	β		
uvbyβ	3500	4100	4700	5500	4860		

Source: From Mihalas and Binney 1981 and Allen 1973.
Wavelength is in angstroms.

3.7 Colors

A hot star has stronger emission in the blue than in the red, whereas a cool star has stronger emission in the red than in the blue, in accordance with Wien's law. As mentioned earlier, brighter objects have smaller magnitudes than fainter objects, so the magnitude in the B band will be a smaller number than in the R band for the hot star. The blue magnitude m_B is often abbreviated simply as B, and similarly for other filters.

The difference in magnitudes between two filters is known as the *color*, or the *color index*: for example, (U-B) or (B-V); the shorter wavelength is listed first. By convention, A0 stars have color indices of 0 for all combinations: that is, (U-B) = (B-V) = (R-I), etc., = 0. So, (U-B) is negative for a very hot star (since there is more ultraviolet than blue emission) and positive for a cool star; the same is true for any of the other color indices. Star colors are listed in Table 3.3 for different spectral types.

As we will explore in Chapters 4 and 5, because of these color variations, different types of stars are highlighted when a galaxy is observed through different filters.

TABLE 3.3 *Colors of main sequence stars*

Type	M_V	$(U\text{-}B)_o$	$(B\text{-}V)_o$	V-R	R-I	V-J	V-K	V-L	V-N
O5	−5.7	−1.15	−0.35						
B0	−4.1	−1.06	−0.31						
A0	0.7	0.00	0.00	0.00	0.00	0.00	0.00	0.00	0.00
F0	2.6	0.07	0.27	0.30	0.17	0.55	0.74	0.8	0.8
G0	4.4	0.05	0.58	0.52	0.41	1.02	1.35	1.5	0.8
K0	5.9	0.47	0.89	0.74	0.66	1.5	2.0	2.5	1.4
M0	9.0	1.28	1.45	1.1	1.1	2.3	3.5	4.3	
M5	11.8	1.2	1.63						

Source: Adapted from Allen 1973, pp. 200, 208.
The subscript o indicates that the color is the star's intrinsic color.

3.8 Hertzsprung-Russell Diagram and Star Clusters

As stars age, their colors and brightnesses change. Color and luminosity variations with time are important to understand in order to interpret the collective starlight in galaxies, which we will examine in Chapter 4. The evolution of stars began to be understood when the astronomers E. Hertzsprung and H. Russell realized that stars' luminosities and temperatures are correlated and fall within certain ranges. The *Hertzsprung-Russell diagram*, or simply H-R diagram, is a plot of the stellar intensity (or, equivalently, the brightness, absolute magnitude, or apparent magnitude) as a function of surface temperature (or spectral type), as shown in Figure 3.13. By convention, the temperature increases to the left, with brighter magnitudes upward.

The *main sequence* is a nearly diagonal line along which most stars are located. Stars are divided into different *Luminosity Classes* (designated by Roman numerals) to signify their location on an H-R diagram: supergiant stars are Luminosity Class I, giants are III, and main sequence stars are V, as indicated in Figure 3.13; intermediate luminosities are labeled classes II and IV. Main sequence stars are sometimes referred to as "dwarfs" because they are below the supergiant and giant branches (this nomenclature is not to be confused with *white dwarfs*, which are old stars that lie below the main sequence).

It is more efficient to collect light through broadband filters than through spectroscopy; thus, determining a star's color is faster in general than determining its spectral type. Equivalent to the H-R diagram is the *color-magnitude diagram* (which plots, for example, B versus B-V), as shown in Figure 3.14. A similar plot is the *color-color diagram* (which plots, for example, U-B versus B-V), as shown in Figure 3.15. In each case, the main sequence occupies a unique line.

Often stars form in clusters. Studies of the H-R diagrams of clusters have led to many insights about the process of stellar evolution. Clusters that contain hundreds of thousands of stars arranged in a spherical, centrally condensed distribution are called *globular clusters*, as shown in Figure 3.16. Clusters that contain only hundreds or thousands of stars arranged in a loose configuration are called *open clusters* or *galactic clusters*, as shown in Figure 3.17. Globular and galactic clusters have different properties and locations, as we will explore in Chapter 8 for the Milky Way.

In many galaxies, we can receive only the integrated light from a cluster. When individual stars can be resolved, we see that the H-R diagrams for open and globular clusters are quite different. In general, globular clusters are much older than galactic clusters, so high mass stars have already evolved off the main sequence.

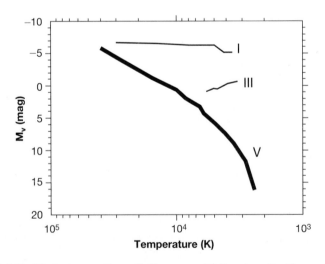

FIGURE 3.13 *Hertzsprung-Russell diagram with Luminosity Classes I, III, and V indicated. The main sequence is the lower dark line labeled V.*

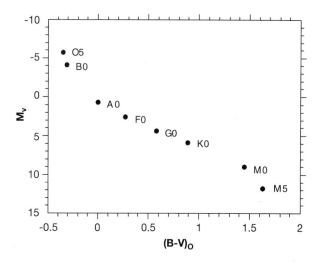

FIGURE 3.14 *Color-magnitude diagram showing the main sequence.*

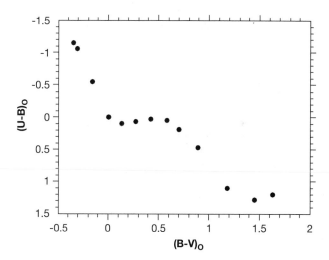

FIGURE 3.15 *Color-color diagram showing the main sequence stars.*

Consequently, globular clusters show a much more prominent giant branch than do younger galactic clusters. The age of the highest mass star still on the main sequence is a measure of the age of the cluster; this point on the main sequence is known as the *turn-off point*. For example, B stars live about 10^8 years. If there are no B stars on the main sequence, the cluster must be at least 10^8 years old. Figure 3.18 shows an H-R diagram for open clusters of different ages.

In the next two chapters, we will make use of our knowledge of star properties to interpret the light observed from different galaxies.

FIGURE 3.16   *(left) A globular cluster, G1 in M31, containing about 300,000 stars. (Imaged with the Hubble Space Telescope WFPC2 by M. Rich, K. Mighell, and J. Neill, Columbia University; W. Freedman, Carnegie Observatories; and NASA.) (right) Globular cluster M3 in our Galaxy. (From the STScI Digital Sky Survey.)*

FIGURE 3.17   *The Jewel Box, NGC 4755, is a galactic cluster in the southern hemisphere of our Galaxy. (Image from the STScI Digital Sky Survey.)*

3.9 Stellar Evolution

Studies of H-R diagrams have led to detailed analyses of stellar structure and evolution. Although a complete review of these processes is beyond the scope of this book, we will briefly summarize the physics behind the key evolutionary stages. A more detailed examination is in references such as Bohm-Vitense 1992.

Stars spend 90% of their lives on the main sequence, while they are converting hydrogen into helium in their cores. A star occupies only one point on the main sequence during its lifetime; this point is determined by the star's mass. During this time on the main sequence, stars are in *hydrostatic equilibrium*, which means that the gravitational forces pulling inward are exactly balanced by

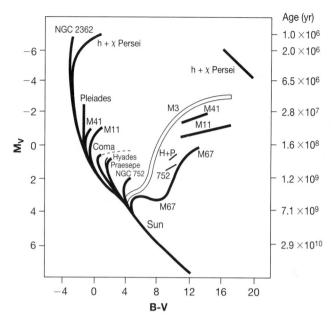

FIGURE 3.18 *H-R diagram for open clusters; the age axis refers to the turn-off point of each cluster. (From Sandage 1970.)*

pressure forces from the thermonuclear reactions pushing outward. This balance of forces may be described by the equation of hydrostatic equilibrium

$$\frac{dP}{dr} = -\frac{Gm(r)}{r^2}\rho(r)$$

for pressure P, radius r, mass at a given radius m(r), and density at that radius $\rho(r)$; G is the gravitational constant (see Appendix 1). Note that the left-hand side divided by the density gives an acceleration, since it is a force per unit volume; the right-hand side is the gravitational force density.

On the main sequence, a star is also in *thermal equilibrium*, because the energy generated by nucleosythesis is radiated away. The core region is where the temperature is at least 10^7 K, which is the minimum (*threshold*) temperature for H-He nucleosynthesis to occur. This high temperature is required in order for protons to overcome their *electrostatic repulsion*; if they move sufficiently fast, then they can come close enough together for the *strong force*, which is 137 times stronger than the electrostatic force at close range, to operate.

The source of a star's radiation is the conversion of mass into energy during nucleosynthesis. The *proton-proton (p-p) chain* and *carbon-nitrogen-oxygen (CNO) cycle* are processes by which low mass and high mass stars convert hydrogen into helium. In both processes, four hydrogen nuclei ultimately are converted into a helium nucleus. During the steps of both nucleosynthesis processes, the conversions are *exothermic*, meaning that they involve the release of energy. The conversion of hydrogen into helium releases an amount of energy equal to 0.007 times the mass of each proton involved, through the familiar relation $E = mc^2$.

In low mass stars, the proton-proton chain is the dominant fusion sequence. In this process, two protons combine to form a proton and a neutron, giving off a *positron* (a positively charged particle with the mass and spin of an electron) and a *neutrino* (a very light or massless particle). The proton and neutron together are the nucleus of *deuterium*, which is a heavy isotope of hydrogen. Next, another proton combines with the deuteron to form a light isotope of helium, plus a gamma ray. Finally, two light helium nuclei combine to form a helium nucleus which has two protons and two neutrons, and two hydrogen nuclei as well. The chain can be summarized as follows:

$$^1H + {}^1H \rightarrow {}^2H + e^+ + \nu$$
$$^1H + {}^2H \rightarrow {}^3He + \gamma$$
$$^3He + {}^3He \rightarrow {}^4He + 2\,{}^1H$$

This process produces 85% of the energy in a solar-type star; there is a net release of 26.2 MeV for each helium nucleus formed. Another branch of the p-p chain combines light and normal helium nuclei to produce beryllium, which forms lithium; another hydrogen nucleus then leads to the formation of two helium nuclei. This chain accounts for nearly 15% of the energy output. Two other p-p branches supply the remaining energy output.

The CNO cycle requires a threshold temperature of 1.6×10^7 K, so it occurs efficiently only in the cores of high-mass main sequence stars. In these cores, three nuclei of helium atoms (*alpha particles*) can combine to form the nuclei of carbon in the *triple-alpha process*. In the CNO cycle, additions of protons to carbon nuclei lead to nitrogen formation, and another proton combines to form oxygen. An unstable isotope of oxygen then decays to form nitrogen, which combines with a proton to yield carbon and helium. These steps can be summarized as follows:

$$^{12}C + {}^1H \rightarrow {}^{13}N + \gamma$$
$$^{13}N \rightarrow {}^{13}C + e^+ + \nu$$
$$^{13}C + {}^1H \rightarrow {}^{14}N + \gamma$$
$$^{14}N + {}^1H \rightarrow {}^{15}O + \gamma$$
$$^{15}O \rightarrow {}^{15}N + e^+ + \nu$$
$$^{15}N + {}^1H \rightarrow {}^{12}C + {}^4He$$

This cycle generates about 25 MeV for each helium nucleus formed.

When all of the core hydrogen has been depleted, a star contracts because it no longer can exert an outward pressure to counteract gravity. The star heats up as gravitational energy is converted to thermal energy and then expands in the outer layers and cools at the surface. The star becomes redder; it is called a *red giant* or *supergiant* at this stage, depending on whether it has a mass less than or greater than about 10 M_o, respectively. (M_o is one solar mass; see Appendix 2.) As a star evolves, its luminosity varies by many orders of magnitude. The core eventually heats up to about 10^8 K, which is sufficient for helium to convert into carbon and oxygen.

This temperature threshold occurs suddenly in a *helium flash* for low mass stars, because the gas just before the flash is *degenerate*. In an ideal gas, the average electron energy is given by the thermal energy, $E = 3kT/2$. In a degenerate gas, the density is so high that electrons fill every available energy state; then the average energy of the electrons is higher than the thermal energy. Degeneracy does not occur in high mass stars at this stage because they have already achieved the threshold temperature for helium burning while on the main sequence.

In the shell immediately surrounding the core, temperatures are now in excess of 10^7 K, so hydrogen nucleosynthesis occurs. This process is called *shell burning*. In high mass stars, there are many different shells of nucleosynthesis, because the core collapse causes more and more of the surrounding layers to heat up and undergo nucleosynthesis. A high mass star resembles an onion at this stage, with hydrogen shell burning in the layer with a temperature of 10^7 K that is farthest from the core, helium burning in the shell that is 1.6×10^7 K, carbon burning closer in, and so on. This sequence is shown in Figure 3.19.

The equilibrium that exists in a star on the main sequence is disrupted as the star enters its giant or supergiant stage. While trying to achieve a new balance of outward pressure and inward gravity during the core contraction, a star may oscillate in radius for thousands of years because of unstable outer layers. The instability in some stars results from a change in the opacity of the layers at certain pressures and temperatures, because of the ionization of helium. If a star expands too much, the outlying layers do not have enough support, so they collapse to smaller radii; the collapse causes them to heat up and expand again.

Stars that oscillate in this stage are called *variable stars* because, as their radius changes, their luminosity changes. There are many different types of variables, which are classified according to the duration and magnitude of their *light curve*, or light variation with time. The time scale for the radial oscillations is typically on the order of days. *Cepheid* variables are stars with masses up to about 6 solar masses, whereas *RR Lyrae* variables are lower mass. Henrietta Leavitt discovered that the period of the oscillation for Cepheids is correlated with the star's absolute luminosity: intrinsically brighter stars have longer periods. The significance of this *period-luminosity* relation lies in being able to determine distances to distant stars, even in other galaxies. We will return to this discussion in Chapter 12, when we consider different distance indicators.

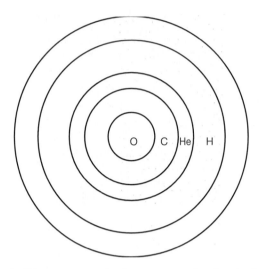

FIGURE 3.19   *Shell burning in a high mass star; the chemical elements shown are undergoing nucleosynthesis in each layer.*

At the latter stages of a star's life, it runs out of fuel that it can process. Low mass stars never get hot enough to fuse oxygen, so when the core is just carbon, nitrogen, and oxygen, nucleosynthesis stops. The end result is a *white dwarf*, which is again a degenerate gas. Its maximum mass is $1.44\ M_o$, which is the amount that can be supported by the pressure of the degenerate gas. This limiting mass is called the *Chandrasekhar limit* in honor of the astrophysicist who developed the idea.

The outer layers of the star gradually detach themselves from the white dwarf core, and during this time the star is said to be a *planetary nebula*, as shown in Figure 3.20. These outer layers mix with the rest of the gas between the stars over periods of millions of years. We can detect planetary nebulae in other galaxies by their emission lines.

Higher mass stars, in contrast, can fuse nuclei all the way up to the iron atom, whose nucleus is the most stable of all the elements. It takes energy to process iron, however, so at this stage core nucleosynthesis stops. In the absence of radiation pressure from nuclear fusion, the core of the high mass star collapses against gravitational forces. A shock wave is produced during the implosion, which leads to an explosion outward at about 2000 km s^{-1} as the star becomes a *supernova*. The outer 80% of the star's layers undergo *explosive nucleosynthesis*, producing all the heavy elements up through uranium during the milliseconds of the detonation and for the next several days afterward. The core collapses further to become a neutron star, because the mass exceeds the Chandrasekhar limit; the electrons and protons are crushed together into a sea of neutrons. Supernovae are studied in other galaxies through their light curves and emission lines; a recent supernova in the Large Magellanic Cloud is shown in Figure 3.21.

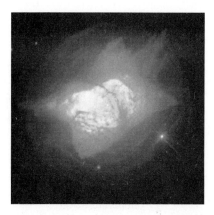

FIGURE 3.20 *Planetary nebula NGC 7027. Faint concentric shells in the outer regions are the ejected star layers. (Imaged with the Hubble Space Telescope WFPC2 by H. Bond, STScI, and NASA.)*

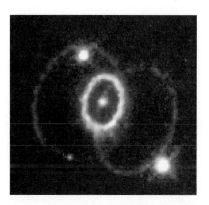

FIGURE 3.21 *Supernova 1987A in the Large Magellanic Cloud, with rings of emission. The rings probably are in different planes, and the outermost ones may be the ends of conical structures formed by high energy beams of radiation. (Imaged with the Hubble Space Telescope WFPC2 by C. Burrows, European Space Agency/STScI, and NASA.)*

Theoretical evolutionary paths for stars of different masses are shown on an H-R diagram in Figure 3.22, and Figure 3.23 shows a computer model of details in the evolution of a 5 M_o star.

Main sequence red stars have ages of several billion years, but red supergiants (which evolve from faster-burning high mass stars) are only 10^8 years old. This means that color alone does not indicate the age of a star or group of stars. The color evolution of ensembles of stars is explored further in Chapter 4.

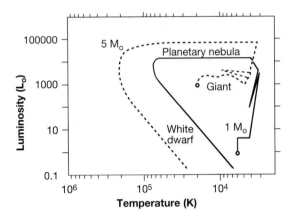

FIGURE 3.22 *Evolution of 1 M_o and 5 M_o stars on the Hertzsprung-Russell diagram. The stars begin on the main sequence (indicated by two small dots), then proceed to the giant stage, planetary nebulae, and white dwarfs. Even higher mass stars become supergiants and subsequently neutron stars.*

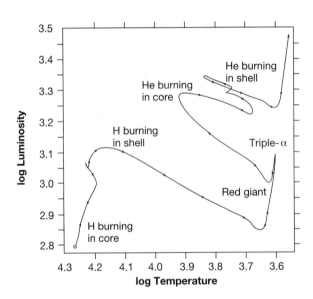

FIGURE 3.23 *Theoretical evolutionary track for a 5 M_o star. The star begins in the lower left on the main sequence and proceeds to cooler surface temperatures. (Adapted from Iben 1991.)*

Exercises

1. What is the approximate spectral type of a star whose peak wavelength is 7000 Å?
2. Star A is a B5V star with $m_V = 5.9$. Star B is a B5V star with $m_V = 7.3$. How do their absolute luminosities compare? What are their relative distances?

3. From the web, import Larry Marschall's stellar photometry lab from Project CLEA (Contemporary Laboratory Experiences in Astronomy): http://www.gettysburg.edu/project/physics/clea/CLEA/home.html, or through ftp via ftp io.cc.gettysburg.edu or ftp 138.234.4.10. This exercise allows you to pretend to observe a cluster of stars with a photometer in order to produce a color-magnitude diagram.

4. What is the approximate spectral type of a main sequence star 100 pc away with an apparent visual magnitude of 9 (assume that no dust is in the way)?

5. A star has a color index B-V = -0.2. Approximately what are its temperature and spectral type?

6. Assume that two identical regions of pure hydrogen have temperatures of 7000 K and 6000 K. How do the populations of level 2 compare with the populations of level 1 for the two regions?

7. How much bigger is the orbit of an electron that has just radiated a Balmer line than the orbit of one that has just radiated a Lyman line?

8. The new Vassar 32-in. telescope sees objects out to about 18th magnitude. How much fainter are these than objects our unaided eye can see?

9. Determine the approximate radii of a main sequence, giant, and supergiant star, all with surface temperatures of 4000 K.

Computer simulation exercises on stellar atmospheres, stellar evolution, and H-R diagrams are available in J. Danby, R. Kouzes, and C. Whitney, *Astrophysics Simulations: Consortium for Upper Level Physics Software* ed. R. Ehrlich, W. MacDonald, and M. Dworzecka (New York: Wiley & Sons, 1995).

Unsolved Problems

1. How accurately do we know the evolutionary tracks of stars?
2. What factors determine the formation of clusters of stars?

Useful Websites

Catalogs and atlases of stellar spectra and synthetic spectra are available through
http://adswww.harvard.edu/ads_catalogs.html
http://www-astro.phast.umass.edu/data/spectra.html
http://zebu.uoregon.edu/spectra.html

Project CLEA has information on retrieving stellar spectra from NASA archival data:
http://www.gettysburg.edu/project/physics/clea/CLEA/home.html

HST archival data are available through http://archive.stsci.edu

Cluster images are available through http://seds.lpl.arizona.edu:80/messier/index.html

Further Reading

Allen, C.W. 1973. *Astrophysical quantities*. London: Athlone Press.

Bessell, M., and G. Stringfellow. 1993. The faint end of the stellar luminosity function. *Annual Review of Astronomy and Astrophysics*, 31:433.

Bohm-Vitense, E. 1992. *Introduction to stellar astrophysics*, vols. 1–3. Cambridge: Cambridge University Press.

Buser, R., and R. Kurucz. 1992. A library of theoretical stellar flux spectra. I. Synthetic UBVRI photometry and the metallicity scale for F- to K-type stars. *Astronomy and Astrophysics* 264:557.

Chiosi, C., G. Bertelli, and A. Bressan. 1992. New developments in understanding the HR diagram. *Annual Review of Astronomy and Astrophysics* 30:235.

Iben, I. 1991. Single and binary star evolution. *Astrophysical Journal (Suppl.)* 76:55.

Jimenez, R., and J. MacDonald. 1996. Stellar evolutionary tracks for low-mass stars. *Monthly Notices of the Royal Astronomical Society* 264:557.

Kaler, J.B. 1989. *Stars and their spectra: An introduction to the spectral sequence.* Cambridge: Cambridge University Press.

Lang, K. 1974. *Astrophysical formulae: A compendium for the physicist and astrophysicist.* New York: Springer-Verlag.

Maeder, A., and P. Conti. 1994. Massive star populations in nearby galaxies. *Annual Review of Astronomy and Astrophysics* 32:227.

Malin, D., and P. Murdin. 1984. *Colours of the stars.* Cambridge: Cambridge University Press.

Mihalas, D., and J. Binney. 1981. *Galactic astronomy: Structure and kinematics.* San Francisco: W.H. Freeman.

Renzini, A., and F. Pecci. 1988. Tests of evolutionary sequences using color-magnitude diagrams of globular clusters. *Annual Review of Astronomy and Astrophysics* 26:199.

Sandage, A. 1970. Main-sequence photometry, color-magnitude diagrams, and ages for the globular clusters M3, M13, M15, and M92. *Astrophysical Journal* 162:841.

Schwarzschild, M. 1958. *Structure and evolution of the stars.* New York: Dover.

CHAPTER 4

Integrated Galaxy Light

Chapter Objectives: to describe measurements and interpretations of the total light from galaxies

---

Toolbox:

absorption	galaxy colors
reddening	surface brightness
equation of transfer	diameter
Luminosity Class	stellar mass function
Yerkes class	population synthesis

4.1 Integrated Light

The total flux of light from a galaxy provides information about the numbers and types of stars that are present. We refer to this light as *integrated light,* in the sense that it is the light from the entire galaxy. The integrated light is measured as a function of passband, obtained through photometry. We will examine how it varies as a function of galaxy type and consider the differences in terms of models of starlight.

In order to determine the true integrated light from a galaxy in different passbands, we must correct the observed light for several effects, including the extinction of starlight by dust in our galaxy and in the disks of other galaxies and Doppler-shifting of the starlight due to galaxy motions. In addition, any galaxy observation is a combination of galaxy plus sky, so careful sky subtraction is essential. Once these corrections are made, the observed integrated light and colors (difference in integrated light between two passbands) of a galaxy can be used for detailed modeling of its constituents.

The *total magnitude* is a measure of the integrated light from the entire galaxy, measured in a particular passband. It is difficult in practice to measure directly the total magnitude of a galaxy because the light is faint in the outer regions and blends into the sky background.

The *metric magnitude* is the integrated light measured within an aperture. The goal of observing is to select an aperture size that includes nearly the entire light from the galaxy but is not so much larger that it also includes a lot of sky. Photoelectric photometers are light collectors in which an aperture of a given size can be selected at the telescope during the observation. With CCD imaging, apertures can be selected after the observation, at the computer. In comparisons of different Hubble types, metric magnitudes may not be appropriate because the galaxies may have different light distributions. An aperture selected to be close to the apparent size of a galaxy may include most of the flux from one galaxy but miss faint extended outer parts of another galaxy, because the edge of a galaxy is sometimes difficult to determine.

The integrated light of a galaxy can also be measured out to some limiting *surface brightness*. Surface brightness is the luminosity divided by the area, commonly expressed in magnitudes per square arcsecond (see Section 4.12). The integrated light measured in magnitudes out to some surface brightness is known as the *isophotal magnitude*, and the radius is known as the isophotal radius. Usually, the isophotal magnitude is the integrated light in a given passband measured out to R_{25}, the radius at which the surface brightness is 25 mag arcsec^{-2}.

The distance modulus is used to convert observed apparent magnitude into absolute magnitude. (A discussion of distance determinations is given in Chapter 12.) Typical elliptical, lenticular, and spiral galaxies earlier than type Sc in Tully's *Nearby Galaxies Catalog* have absolute blue magnitudes $-22 < M_B < -16$, whereas later-type spirals and Magellanic irregulars (measured in the Local Group) have $-18 < M_B < -12$, often fainter than -16 mag. The brightest elliptical galaxies have $M_B \sim -22$; in contrast, the faintest known dwarf E in the Local Group has $M_B \sim -9$, which is only as bright as a globular cluster.

Now let us consider the corrections that should be made to observations of a galaxy to determine its true integrated light in a given passband: *extinction* from our Galaxy along the line of sight to the galaxy, internal extinction in the observed galaxy, and a correction due to the galaxy's motion with respect to ours, which gives rise to a redshift. Extinction is a combination of scattering and absorption, which cause dimming and reddening of starlight. The galaxy's redshift causes light to be shifted toward longer wavelengths. The entry B_T in the RC2 and RC3 is the apparent total blue magnitude, measured from extrapolations using successively larger photometric apertures. The entry B_T° is the corrected total blue magnitude. Therefore B_T° is given by

$$B_T^\circ = B_T - A_B(\text{Milky Way}) - A_B(\text{internal}) - K_B$$

for Milky Way absorption A_B(Milky Way), internal absorption A_B(internal), and redshift correction K_B, all determined for B band here. The details of these terms are discussed in the following sections.

4.2 Milky Way Absorption

The disk of our Galaxy is filled with a mixture of gas and dust. To a first approximation, this *interstellar matter* is distributed uniformly in the midplane of the galaxy. As we will see in Chapter 6, it is mostly concentrated within 100 pc of the midplane. To view another galaxy, we must look through the dust, which blocks and reddens that light. The amount of dust depends on the location of the observed galaxy with respect to our disk. Galactic latitude (b) is the most important factor in determining the amount of dust along a line of sight, since objects at higher latitudes are viewed through less Milky Way dust; see Figure 4.1.

The absorption in blue light perpendicular to the disk—that is, at galactic latitude b = 90°—has a value $A_{B,0}$ of about 0.5 mag (according to de Vaucouleurs, although this value may be too high according to other observers); sometimes the term Δm_0 is used to represent this absorption. The absorption along a line of sight at an arbitrary latitude b is therefore given by

$$\Delta m_b = \Delta m_0 \csc b$$

Second-order variations in absorption as a function of both latitude and longitude are caused by the patchy distribution of spiral arms, dust lanes, and clouds in the midplane of the Milky Way. Position-dependent absorption has been determined in our Galaxy on the basis of variations of galaxy counts and star colors along different lines of sight; a simple model as a function of latitude and longitude is tabulated in the RC2 and RC3 on the basis of the analyses of several different observing teams. In Chapter 6, we will return to this subject and consider the correlation of hydrogen gas emission and galaxy counts with reddening, which leads to one of the best mappings of absorption in the Milky Way.

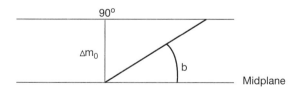

FIGURE 4.1 *Absorption caused by Milky Way dust along a line of sight.*

4.3 Radiative Transfer

A term commonly used instead of absorption is *optical depth*, τ, which is a dimensionless quantity related to the blockage of light; it is related to the number and properties of the intervening dust grains. Dust grains can absorb photons. Absorption causes the dust grains to heat up; they then cool off by reradiating infrared photons in random directions. Alternatively, grains can scatter photons. Because only a few of these photons are scattered back into our line of sight, these processes have the effect of removing light from our view. The processes of absorbing and scattering light are collectively known as *extinction*. The amount of extinction along a line of sight is the sum of the extinctions caused by all the grains on the path. The individual grain extinction depends on the efficiency with which that dust grain absorbs or scatters light of a specific wavelength, which in turn depends on the composition, size, and shape of the particle. The extinction due to all similar dust grains along a line of sight therefore is related to their number density (the number of grains in a given volume) and the path length. For simplicity, let us assume that there are n_d dust particles per cubic centimeter, that each particle has an effective absorption cross section σ, and that we are observing an object through a path of length L, as in Figure 4.2. Then the optical depth, τ, for a given wavelength is given by

$$\tau_\lambda = n_d \sigma_\lambda L$$

The effective cross section σ may be described in terms of the *efficiency* Q_λ, which depends on the properties of the dust grain, and the dust grain geometric cross section, πa^2, for a dust grain of average radius a (for optical extinction, the grains are commonly submicron-size).

That is,

$$\sigma_\lambda = Q_\lambda \pi a^2$$

In practice, it is more realistic to consider several types of grains along a given path and a variable density along the path. Then the optical depth along that line

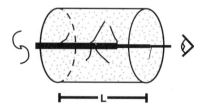

FIGURE 4.2   *Viewing an object through dust, which absorbs and scatters light.*

of sight is the sum of the optical depths due to the different sizes and compositions of the grains, integrated over the path length:

$$\tau = \int_0^L n_1 \sigma_1 dl + \int_0^L n_2 \sigma_2 dl + \int_0^L n_3 \sigma_3 dl + \ldots$$

In even more detail, each grain type actually has a distribution of sizes, whose average we have represented by a; the total opacity is more accurately given by integrating grains of each composition over their spectrum of sizes and then summing over each grain type.

The optical depth can be directly related to the decrease in intensity of light, dI, and the incident intensity, I:

$$dI = -I \, d\tau, \text{ or } \frac{dI}{d\tau} = -I$$

The intensity is constant if there is no absorption. This relation is the simplest form of the *equation of radiative transfer*, which can be integrated to yield

$$\boxed{I = I_o e^{-\tau}}$$

for incident intensity I_o and observed intensity I at a given wavelength. When we discuss gas in Chapter 6, we will consider a more general form of the equation of transfer that includes a term for emissivity as well as extinction.

Because magnitudes of absorption are related to intensity by the logarithm to the base 10 (that is, $A_\lambda = -2.5 \log I/I_o$, as we saw in Chapter 3), whereas optical depth is related to intensity by the natural logarithm, or logarithm to the base e (that is, $\tau = -\ln I/I_o$), we can relate the absorption to the optical depth as

$$A = 2.5 \log e^\tau = 2.5\tau \log e = 2.5(0.434)\tau$$

so

$$\boxed{A_\lambda = 1.086\tau_\lambda}$$

4.4 Reddening

Another complication from dust is that it selectively absorbs or scatters shorter wavelengths more effectively than longer wavelengths. At optical wavelengths, the extinction A_λ is nearly linear with wavelength λ and scales inversely:

$A_\lambda \propto 1/\lambda$. This *selective extinction* suggests that the wavelength at which a galaxy is observed may make a big difference in the morphology. For example, the absorption in the K band (2.2μ) is only about one-tenth that in the visual: $A_K = 0.09\,A_V$. Dust lanes in galaxies have an absorption that is greater than unity in the B band, typically several magnitudes, so they appear dark. In the K band, however, the dust lanes may not show up at all or only very weakly. Thus, it is very important to understand that galaxy morphology can be dependent upon filter selection. Figures 4.3 and 4.4 show B and I band images of two galaxies. Note that the I band image of M51 is similar to the B band image but smoother due to less obscuration by dust; in NGC 7793 the structure differs considerably in the two passbands, due to different distributions of young and old stars. These differences in morphology will be discussed in more detail in Chapters 5 and 9.

Table 4.1 scales the absorption in different filters (on the Johnson scale) to 1 visual magnitude of absorption, based on a standard reddening law derived from observations in our galaxy. Although the reddening law is applicable in general,

FIGURE 4.3   *M51 shown in B band (left) and I band (right). Note the difference in smoothness of the arms. (Imaged with the Palomar 1.2-m Schmidt telescope by the author.)*

FIGURE 4.4   *NGC 7793 shown in B band (left) and I band (right). Note the lack of clear spiral structure in I. (Photographic images obtained with the Palomar 1.2-m Schmidt telescope by the author.)*

TABLE 4.1 *Absorption as a function of passband (Johnson filters)*

Passband	Absorption (mag)	Passband	Absorption (mag)
U	1.60	I	0.42
B	1.33	J	0.24
V	1.00	H	0.18
R	0.64	K	0.090

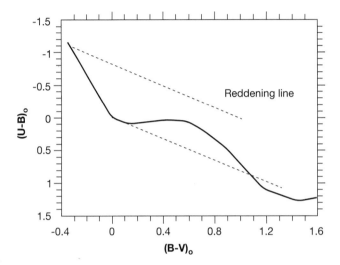

FIGURE 4.5 *Reddening from dust affects star positions on a color-color diagram. The reddening lines indicate the direction of shift of stars from the main sequence line (solid line).*

it may vary occasionally; dust grains may grow to larger-than-average sizes in some molecular clouds, and grains in H II regions may be larger than average due to the depletion of small grains by radiation pressure. To a first approximation, the standard reddening law appears to be the same at optical wavelengths in other disk galaxies; there may be differences in the ultraviolet for some galaxies.

The difference in absorption between two passbands is known as the *reddening*, and also called the *color excess*, $E(\lambda_1 - \lambda_2)$. Note that color excess is also just the difference between the observed color index [e.g., (B-V)] and the intrinsic color index [e.g., $(B\text{-}V)_o$]:

$$E(B - V) = (B - V) - (B - V)_o = A_B - A_V$$

The effect of reddening is to shift individual star positions in a color-color diagram, which was described in Chapter 3. Figure 4.5 shows lines representing the shifts resulting from the reddening of starlight by dust. The slope of the reddening line is a function of the composition and structure of dust grains and is uniform (to first order) across our Galaxy and in other galaxies. It is evident that early-type stars are still recognizable as such, since they are shifted to positions that cannot be confused with other star types.

At the dip in the color-color diagram, which corresponds to the location of A stars, ambiguity sets in. For example, a star on the lower dashed line at $(B-V)_o = 1.2$ could be either a slightly reddened late-type star or a more heavily reddened earlier-type star. In these cases, it is not always possible to sort out the amount of reddening that is present in order to determine star type. Radio observations can help lead to an estimate of the amount of gas and therefore the dust present, as described in Chapter 6.

Dust in our Galaxy also complicates the determination of the distance to an object, whether it is a star in the Milky Way or in a distant galaxy; because dust dims starlight, objects appear to be farther away than they really are. The distance modulus equation, discussed in Chapter 3, is modified by the addition of the extinction term:

$$m_\lambda - M_\lambda = 5 \log r - 5 + A_\lambda$$

When the amount of intervening dust is unknown, distances can be in error by large amounts.

4.5 Internal Absorption

When galaxies are inclined to our line of sight, they do not appear as bright as when they are face-on, because we are looking through their dust layers. Internal galaxy extinction can be derived by comparing galaxies with different inclinations and fitting a model for dust; the results are Hubble-type-dependent, since different types of galaxies have different amounts of dust. In the absence of extinction, there would be no inclination effect. Figure 4.6 shows the observed extinction in Sc galaxies as a function of their inclination.

The RC2 and RC3 make use of the following empirical fit:

$$A_B(i) = 0.70 \log[\sec(i)]$$

for inclination i and blue extinction A_B. Again note that $i = 0°$ means face-on and $i = 90°$ means edge-on. Thus, there is a secant in this equation, rather than a cosecant, which appeared in the equation for Milky Way absorption, because the Milky Way angle is measured from the midplane rather than from the perpendicular. A recent I band study of Sc galaxies by R. Giovanelli and colleagues find the relation

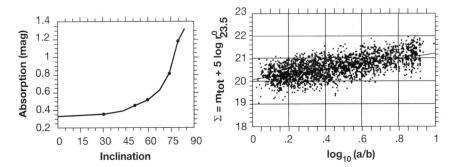

FIGURE 4.6 *(left) The internal visual absorption in an Sc galaxy is plotted as a function of its inclination to our line of sight. (Based on Holmberg's classic (1958) photographic data.) (right) The absorption for Sc galaxies. (Based on I band observations by Giovanelli et al. 1994.)*

$$A_I = 1.12(\pm 0.05)\log\left(\frac{a}{b}\right)$$

for major axis a and minor axis b.

A value of 0.5–1.0 mag of visual extinction is typical through a disk, although some galaxies may be more transparent than this. Studies using galaxies that overlap on our line of sight can provide some insight into disk transparencies. Sometimes arms from the more distant galaxy can be seen through the disk of the intervening galaxy, but sometimes the background galaxy is obscured. If the optical depth is greater than unity, corresponding to about 1 magnitude of extinction, the object is said to be *optically thick*, wheareas with an optical depth <<1 the object is said to be *optically thin*.

4.6 K Correction

Distant galaxies are receding from us more quickly than are nearby galaxies, as demonstrated by Hubble through the relation now known as the *Hubble law*

$$\boxed{v = H_o d}$$

for velocity of recession v in km s^{-1}, Hubble constant H_o = 50–100 km s^{-1} Mpc^{-1} (HST observations indicate a value of about 75 km s^{-1} Mpc^{-1}), and distance d in megaparsecs. Figure 4.7 illustrates the Hubble law; we will return to a discussion of it in Chapter 12. As a result of increasing galaxy recessional velocity with increasing distance, very distant galaxies appear highly redshifted. Because all wave-

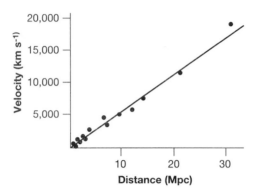

FIGURE 4.7   *The Hubble diagram shows the relation between a galaxy's recessional velocity and its distance. The relation noticeably deviates from a straight line beyond the limits shown here because of the deceleration of the Universe (see Chapter 12). This diagram, made by Hubble and Humason, was published in 1931 in the Astrophysical Journal (volume 74, page 43); the slope has undergone extensive revisions since then.*

lengths are shifted, their light distributions (blackbody curves) are shifted toward longer wavelengths. The distributions are also "stretched," since $\Delta\lambda \propto \lambda\, v/c$.

Redshift effects should be borne in mind when comparing the integrated light of two galaxies in a given passband: Two galaxies at different distances with the same intrinsic brightness in blue light will have different apparent blue magnitudes not only because of the inverse square law, whereby distant objects are fainter, but also because of their recessional velocities due to the expanding Universe, whereby distant objects are redder. The shift of radiation toward longer wavelengths due to recession can be corrected using what is known as the *K correction*.

The *cosmological redshift*, z, relates the rest wavelength, λ, to the difference between the observed wavelength and the rest wavelength, $\Delta\lambda$; the shifted wavelength results from the relative velocity v of the object:

$$z = \frac{\Delta\lambda}{\lambda} = \sqrt{\frac{1 + v/c}{1 - v/c}} - 1$$

This equation reduces to the form z = v/c for nonrelativistic velocities (v << c). Galaxy recession therefore has the consequence of shifting the intensity of the observed energy distribution from what the galaxy emits at each λ to $\lambda(1+z)$. For example, consider a blackbody at 5500 K, zero redshift, which peaks near 5600 Å. The same blackbody with a redshift z = 0.5 shifts to a peak near 8400 Å. As a result, V band light is shifted to the I band, as shown in Figure 4.8.

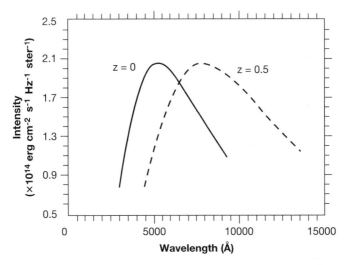

FIGURE 4.8 *A blackbody at a temperature of 5500 K is shown with no redshift (solid line) and a redshift of 0.5 (dashed line).*

The corrections for redshift of the spectra of galaxies depend on Hubble type, because different kinds of galaxies have different stellar mixes and therefore different integrated light distributions. Observational details of their determination are described in the RC2 and RC3. The K correction can be written as $K_B(z,T)$ for the B band, because it is a function of redshift z and Hubble type T. It can be broken into a Hubble type part and a redshift part, $K_B(z,T) = K'_B(T)cz$, where cz = v (in the nonrelativistic case) and $K'_B(T)$ is an empirically determined function of the Hubble type, as listed in Table 4.2.

For example, suppose that v = 10^4 km s^{-1} (corresponding to a redshift z = 0.03) for an Sa galaxy (T = 1). Then the K correction is

$$K'_B cz = [(0.15 - 0.025) \times 10^{-4}] \times 10^4 = 0.125 \text{ mag}$$

which is relatively small. Recall that this term is subtracted from the apparent magnitude to provide a redshift correction. That is, this galaxy is 0.125 mag brighter in the B band than it appears to be after the spectrum has been redshifted.

TABLE 4.2 *K correction terms for different Hubble types*

Hubble type	$K'_B(T)$
T < 0 (elliptical)	0.15×10^{-4}
0 < T < 3 (S0, early type)	$(0.15 - 0.025\,T) \times 10^{-4}$
T > 3 (intermediate, late spiral)	$[0.075 - 0.10(T - 3)] \times 10^{-4}$

Source: From Pence 1976, and listed in the RC2.

For more distant (high z) galaxies, there are also evolutionary effects to be addressed, which require further corrections. Consider, for example, recent HST observations showing that very young galaxies may have had a higher star formation rate than they do now. This could make them bluer than their more evolved (low z) forms; that is, these high redshift galaxies do not have the same spectrum as their evolved counterparts. If a substantial fraction of their light is in the ultraviolet part of the spectrum, then highly redshifted galaxies would be measured as brighter (rather than fainter) in the B band than their intrinsic brightness. The corrected apparent blue magnitude would then be fainter than what is observed, rather than brighter. Such high-z corrections require an evaluation based on detailed models for star formation histories in different types of galaxies.

4.7 Luminosity Class

Now that we have defined a galaxy's magnitude, it is useful to introduce another classification system that is related to absolute magnitude. The *Luminosity Class* of a galaxy is based on a classification system developed by Sidney van den Bergh, which is called the *DDO system* after the David Dunlop Observatory of the University of Toronto, where he was working when he developed the system. This scheme assigns galaxies to various classes similar to Hubble type, but in addition assesses the relative brightness and definition of spiral arms. The galaxies with the best defined and brightest arms are Luminosity Class I; the faintest and fuzziest are Luminosity Class V. Examples are shown in Figure 4.9. Most early Hubble types are Luminosity Classes I–III; later types and irregulars are Luminosity Classes III–V. Subdivisions between the main classes can be represented by hyphens, such as I–II, or by decimals, such as III.4.

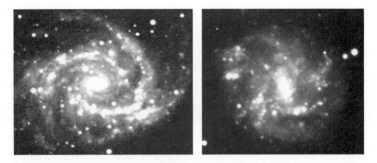

FIGURE 4.9   *NGC 2997 (left) is a Luminosity Class I.3 galaxy. Note the prominent spiral arms. NGC 2500 (right) is a Luminosity Class II.8 galaxy; its arms are fuzzy. (B band CCD images obtained with the KPNO Burrell-Schmidt and 0.9-m telescopes, respectively, by the author.)*

This system, like the Hubble system, is a subjective classification based only on single blue light photographs, but it too turns out to be physically significant, because a galaxy's Luminosity Class is correlated with its absolute magnitude. For Sb and Sc galaxies, Luminosity Classes I–V correspond approximately to absolute blue magnitudes -21 through -16, respectively, but with a large intrinsic rms dispersion of about 0.7 mag. Because of this dispersion, luminosities of spiral galaxies can be estimated only to about a factor of 2 with this method. Thus, Luminosity Class is useful as a crude distance indicator: the Luminosity Class, listed in Sandage and Tammann's *Revised Shapley-Ames Catalog,* gives an indication of the absolute magnitude, and the apparent magnitude can be measured. Then the apparent and absolute magnitudes can be subtracted to give the distance modulus and determine a distance.

4.8 Broadband Colors of Galaxies

Galaxy colors, like star colors, are defined as the difference in magnitude between the integrated light in two different passbands. Broadband photometry reveals different colors for different Hubble types. Table 4.3 is based on several hundred galaxy observations in UBV on the Johnson-Morgan photometric system.

To interpret these observations, recall that smaller magnitudes correspond to bluer objects, because by convention the smaller wavelength comes first in the color index. We see from the table that later types are bluer, as expected on the basis of our discussions of integrated spectral types. Note that E and S0 have about the same colors, which can be understood in terms of similar star formation rates; these will be discussed in Chapter 10. For E and S0 galaxies, the color $(B-V)_o$ rises by 0.25 mag between absolute blue magnitudes $M = -16$ and -18, meaning that the brighter galaxies are redder. This trend is known as a *color-magnitude effect* and probably results from a dependence of mean metallicities on absolute magnitude; that is, it suggests that more-luminous galaxies have more metals and so are redder (as we show in Chapter 6, more blue light is

TABLE 4.3   *Colors for different Hubble types*

Type	$(U-B)_o$	$(B-V)_o$	Integrated spectral type
E	0.50	0.92	K2
S0	0.48	0.92	K2
Sa	0.28	0.82	G9
Sb	0.27	0.81	G8
Sbc	-0.02	0.63	G2
Sc	-0.12	0.52	F8
Im	-0.20	0.50	F7

Source: de Vaucouleurs data, adapted from Vorontsov-Vel'yaminov 1987.

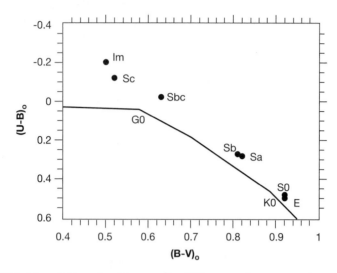

FIGURE 4.10 *Color-color diagram for different galaxy types. Increasing blueness is upward on the ordinate and to the left on the abscissa. The solid line shows main sequence stars for comparison.*

absorbed if there is a higher metallicity). Figure 4.10 shows the locations of these galaxy types on a color-color diagram. Elliptical and early-type spirals are much redder than late-type spirals and irregulars. A knowledge of starlight can be applied to an interpretation of the colors of galaxies. A comparison of this diagram with the color-color diagram for stars (see Chapter 3), indicated by a solid line in the figure, shows the approximate integrated stellar spectral type of each galaxy.

Table 4.4 lists colors in the Johnson photometric system for several well-studied galaxies, ordered by galaxy type; (B-H) is approximate, because the apertures used to measure B and H magnitudes were not the same.

TABLE 4.4 *Galaxy colors*

NGC	Type	(B-V)	(U-B)	(B-H)
3379	E0	0.94	0.52	
4261	E2	0.99	0.55	3.37
4387	E7	0.94	0.56	2.93
4270	S0	0.90	0.41	3.14
3623	Sa	0.90	0.41	2.81
2841	Sb	0.85	0.41	2.93
1300	Sbc	0.68	0.13	2.16
598	Scd	0.55	−0.10	1.56
7793	Sdm	0.59	−0.10	1.65

Source: Optical from RC2; infrared from Tully 1988.

4.9 Yerkes System

Having discussed integrated galaxy spectral types, we can now introduce another galaxy classification system based on integrated light. Although the parameters of the Hubble classification system are correlated with each other, there is a lot of variation within each type, and confusion can result in classifying some galaxies. The spiral galaxy M51 (see Figure 4.3), for example, is classified as an Sb galaxy; its pitch angle is the same as the average for Sb galaxies, but its bulge-to-disk ratio is smaller than normal for this type. W. W. Morgan at the Yerkes Observatory sought to remove some of the ambiguity of Hubble's classification by developing a single-parameter system that is based on the degree of central concentration of light. His classification system is generally referred to as the *Yerkes system*.

Morgan observed that the central concentration is correlated with the integrated spectral type of a galaxy; that is, the kinds of stars in the galaxy are related to the compactness of their distribution. The galaxy types range from a, with small central bulges and diffuse light distributions, to f, g, and then to k, with the brightest central light concentrations. In this designation, k indicates that the integrated spectral type of the galaxy is approximately like a K star. Such galaxies generally are ellipticals.

In addition, the Yerkes system has a *form-family* designation, which describes the overall shape of the galaxy. These include obvious letters such as S for spiral, E for elliptical, and I for irregular; also B for barred spiral and R for galaxies with rotational symmetry (analogous to S0 galaxies). The designation cD denotes galaxies with compact nuclei such as ellipticals but with extended envelopes, as shown in Figure 4.11. They tend to be giant elliptical galaxies in the centers of dense clusters of galaxies, which is an indication that they might have formed through mergers of other galaxies. This possibility will be explored further in Chapter 12.

FIGURE 4.11 *cD galaxy NGC 4889 in the center of the Coma cluster. (Photographed with the Palomar Schmidt telescope. Image from the STScI Digital Sky Survey.)*

4.10 Population Synthesis

A glance at color images of different types of galaxies reveals several prominent trends: bulges tend to be yellow, bars tend to be yellowish red, and spiral arms tend to be blue. Elliptical galaxies tend to be redder than spirals, and irregulars tend to be bluer. J. D. Wray's *Color Atlas of Galaxies* contains images of about 600 galaxies. In the photographic reproductions, Wray has kept the correct color balance at all brightness levels and has the correct surface brightness from one galaxy to another. In these photos, the color differences from galaxy to galaxy of different Hubble type are clearly seen, as well as the different types of stars that make up the arms, interarm regions, and bulges. The Space Telescope Science Institute website given at the end of the chapter has links to HST color photos of a variety of galaxy types.

A simple model of an evolving stellar system gives insight into the interpretation of color images. An ensemble of stars, such as a galaxy, should progressively redden with time, because stars become cooler as they evolve off the main sequence, and high mass stars evolve rapidly and disappear. Decreasing surface temperature leads to a shift in the star's blackbody curve to a longer peak wavelength, as discussed in Chapter 3. Models of how stars evolve can be used to determine the colors or spectra of an ensemble of stars of different types as they age. This process is known as *population synthesis*.

If star formation ceases in a galaxy, it reddens as its stars age. However, if stars continually form throughout most of a galaxy's lifetime, the increasing redness of the older stars is mixed in with the blueness of the younger high mass stars forming. In order to understand how galaxy light varies with time, let us consider in the next section the numbers of stars of different types and their colors and ages. These taken collectively add up to the light that we see in galaxies.

4.11 Stellar Mass Function

Stars have masses that range from 0.07 M_o to about 50 M_o. Their lifetimes are a function of their mass, since higher mass stars achieve greater internal temperatures and therefore process nuclear fuel more rapidly. The main sequence lifetime, τ_{MS}, is approximately given by:

$$\tau_{MS} \approx 10^{10} \frac{M}{L} \text{ yr}$$

for star mass M in solar masses and luminosity L in solar luminosities. For example, O stars live about 10 million years and M stars live about 20 billion years. Typical parameters for stars are given in Table 4.5.

Higher mass stars are less abundant than lower mass stars; there is a logarithmic decrease in the number for higher mass. The *initial mass function* (IMF) is the

TABLE 4.5 *Star parameters*

Stellar type	Mass (M_o)	Luminosity (L_o) in V	MS lifetime (yr)
O5	40	2.5×10^5	1.6×10^6
B0	16	2.5×10^4	6.4×10^6
A0	4	80	5×10^8
G2	1	1	10^{10}
M0	0.5	0.06	7.9×10^{10}

Source: Luminosity and mass from Mihalas and Binney 1981.

number of stars as a function of mass that form in a region of star formation. It is given by the number of stars per unit mass per unit volume, $N(M) \propto M^{\alpha}$. The exponent α is often represented as -2.35, which is the *Salpeter mass function* value, although sometimes the distribution function is broken up into three discrete regions of low-mass, intermediate-mass, and high-mass stars, with exponent -1.25 for $0.4 < M_o < 1$, -3.2 for $3 < M_o < 20$, and -2.35 for $1 < M_o < 3$, as shown in Figure 4.12. There have been many other fits to the IMF; for example, the *Miller-Scalo mass function* has a Gaussian distribution, where the number of stars per square parsec with a logarithmic mass between log M and d log M, $\xi(\log M)$, is given by

$$\xi(\log M) \propto \exp[-1.08 \log (10M^2)]$$

On the main sequence, the luminosity L and radius R both scale with mass: $L \propto (M)^{3.2}$ (more precisely, the exponent is 2.3 for $M < 0.5\ M_o$ and 4.0 for $M > 0.5\ M_o$), and $R \propto (M)^{0.7}$.

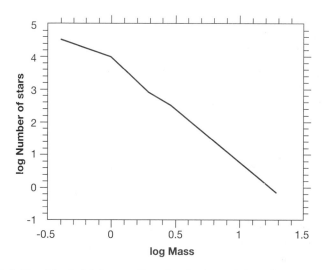

FIGURE 4.12 *The initial mass function is the number of stars as a function of their mass, with different slopes for low-, intermediate-, and high-mass stars.*

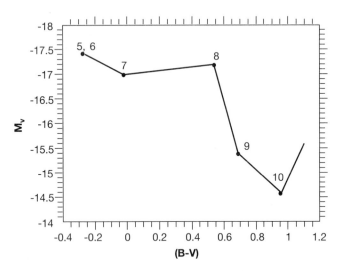

FIGURE 4.13 *Population synthesis model for a Salpeter initial mass function;*
log (age) is indicated in years above the curve. (Based on Wray 1988.)

The initial mass function, sometimes called the birth mass function, might vary as a galaxy evolves, so it is useful to consider the mass function as a time-dependent quantity. In practice, there is no hard evidence that the IMF varies either in time or in place. In Wray's atlas, the time change in color of an ensemble of stars with a Salpeter mass function was determined from stellar evolution tracks. The diagram in Figure 4.13 shows the integrated galaxy color as a function of time. This simple model shows why most disks are yellowish. The ensemble moves from left to right with time in the diagram, showing that it progressively reddens and dims as it ages. A comparison of observations with the model colors indicates that most of the stars in galaxies are more than 10^{10} years old. In addition, there are often young star-forming regions, especially in late-type spiral galaxies and irregular galaxies. Initially, the newly formed stars are still hidden by the dust in which they formed. Individual young blue knots in galaxies are observed to have (B-V) = 0.2, whereas a pure old yellow stellar population has (B-V) ~ 0.95.

The interpretation of galaxy colors is further complicated by the fact that there may be bursts of star formation, which lead to spikes in the mass function at the time when new high-mass (blue) stars are reintroduced into galaxy populations, or there may be continuous star formation going on at a given rate, which also requires new input into the galaxy colors. Furthermore, both age and metallicity cause a galaxy to redden. Detailed evolutionary models that take into account different mass functions and metallicities can be used to sort out these effects. Models available electronically (as of the writing of this book) include those by C. Leitherer and T. Heckman, listed in the websites. Plots showing galaxy colors as a function of age are shown in Figure 4.14.

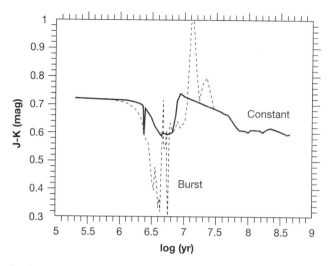

FIGURE 4.14 *Infrared color as a function of age for a cluster in the Leitherer and Heckman (1995) instantaneous burst model (dashed line) and constant star formation model (solid line) with solar metallicity.*

It is instructive to examine the primary source of stellar radiation in different passbands in order to determine how morphology may vary with filter. Light in the B band is dominated by stars only 10^7–10^8 years old. In the I band, the stars that contribute most are the K and M giants, which are evolved solar mass stars whose ages are 10^9–10^{10} yr; thus, the I band traces older stars. Supergiants also make a substantial contribution to the I band star for star, but because they are few in number, their contribution to the total I band light is small. Table 4.8 lists, for different stellar types in the solar neighborhood of the Milky Way, the space

TABLE 4.8 *Intensities for stars*

Type	Space density (number/pc³)	2H (pc)	L_I	I_I/I_{total} (%)	L_V	I_V/I_{total} (%)
OV	2.5e–8	110	3.25e5	0.35	5.01e5	5.98
BV	1.0e–4	120	6.49e2	3.09	794	41.3
AV	5.0e–4	330	151	9.88	19.9	14.2
FV	2.5e–3	380	4.53	1.71	2.51	10.3
GV	6.3e–3	680	1.8	3.06	0.79	14.7
KV	1.0e–2	700	0.705	1.96	0.158	4.8
MV	5.0e–2	700	0.233	3.24	7.94e–3	1.2
K–MIII	6.0e–6	700	4.53e4	75.54	398	7.3
MIa	4.0e–10	120	2.27e6	0.04	1.58e4	3.3e–3
MIb	1.0e–8	120	2.27e6	1.08	3.16e4	0.16

Source: From Elmegreen 1979, and Allen 1973.

density n (number of stars per cubic parsec), scale height, 2H, perpendicular to the disk, and luminosity per star in the I band and in the V band. The percentage of the total light that each type contributes to these bands in the absence of extinction is given by

$$\frac{I}{I_{total}} = \frac{n(2H)L}{\sum_{\text{all types}} [n(2H)L]}$$

Note that K and M giants contribute 75% of the integrated I band light from a galaxy, whereas they contribute only 1% of the V band light. In contrast, B stars contribute only 3% of the I band light but 41% (the dominant contribution) of the V band light.

4.12 Surface Brightness

Surface brightness is a measure of intensity, or flux per unit solid angle (square arcsec), from a galaxy, and is sometimes abbreviated as Σ. This quantity is the same as $1/4\pi$ times the luminosity per unit area of the galaxy. The surface brightness can also be expressed in magnitudes per square arcsecond (mag arcsec^{-2}), as μ, where $\mu \equiv -2.5 \log \Sigma$ +constant. This is analogous to the relationship between magnitudes and luminosities. To determine the surface brightness averaged over an entire galaxy, one must divide the total integrated light by the solid angle of the galaxy. For this purpose, a discussion of galaxy diameters is given in the following section. The concept of surface brightness is useful because intensity is distance-independent; although light flux decreases as the inverse square of distance, the angular area subtended by a source decreases as the square of distance, too. Thus, the flux per unit angular measure remains constant. With sufficient resolution, the variation of surface brightness as a function of position in a galaxy provides important information about the distributions and types of stars across a galaxy disk, as we will discuss in Chapter 5.

There is currently a lot of interest and research on Low Surface Brightness (LSB) galaxies (see Figure 4.15), which have normal or larger-than-normal sizes and masses but whose surface brightnesses are several magnitudes per square arcsecond fainter than normal galaxies; their central surface brightnesses are fainter than 23.0 mag arcsec^{-2}. This low surface brightness results in a lower integrated magnitude out to R_{25}; for example, a low surface brightness dwarf (LSBD) elliptical has $M_B \sim -13$ instead of -16. This research on intrinsically faint galaxies is very important, because many of the established ideas about galaxies are based on a standard brightness. There may be many galaxies that are not detected because of their low surface brightness, and they may have very different properties (such as star formation rates) than the brighter galaxies. Their distribution, history, and evolution may have important cosmological consequences.

FIGURE 4.15 *Low surface brightness (LSB) galaxy UGC 1230, Hubble type Sc. (Image from the Electronic Universe website, by G. Bothun, University of Oregon.)*

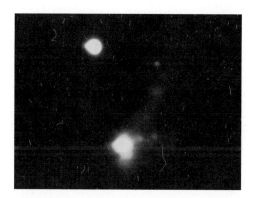

FIGURE 4.16 *Blue Compact Dwarf galaxy IIZw40. (Imaged in B band with the KPNO 0.9-m telescope by J. Salzer, Wesleyan University.)*

In contrast to the LSBs are the *BCD*s, or Blue Compact Dwarfs. These galaxies are brighter than normal for their size and appear to be undergoing a burst of star formation. An example is shown in Figure 4.16, and further discussion is in Chapter 11.

4.13 Diameters

Galaxy diameters may be specified in terms of some limiting brightness level or a morphological criterion such as the ends of spiral arms. Although there are many uncertainties in these methods, disk galaxies appear to fade out rather abruptly and at fairly uniform light levels from one galaxy to the next.

The estimated diameter must be corrected for inclination effects. There are two diameters listed in the RC2 and RC3: One gives the diameter out to a surface brightness of 25 mag arcsec^{-2}, which is also known as the *de Vaucouleurs diameter*. The de Vaucouleurs diameter is listed as log D_{25} in the RC2, where D_{25} is in tenths of arcminutes. Thus, an entry of log $D_{25} = 1.84$ corresponds to a diameter in arcminutes of $10^{1.84}/10 = 6.'9$. The other entry is the diameter corrected for absorption effects due to inclination, listed as log D_o, which is sometimes known as the *standard diameter* or corrected "face-on" diameter. Sometimes galaxies can be measured down to a level of 26.5 mag arcsec^{-2}, which is known as the *Holmberg diameter*, D_H. Recent improvements in detectors and sky subtraction algorithms have made it possible to trace galaxy disks to as faint as 28 or 29 mag arcsec^{-2}.

A conversion from apparent magnitude to surface brightness (magnitudes per square arcsecond) depends on the size of a galaxy. Since the physical quantity involved is the luminosity, it is necessary to divide the luminosity by the area of the galaxy and convert that to surface brightness μ. Let L/A represent the galaxy luminosity L divided by galaxy area A. This can be rewritten as

$$\mu = -2.5 \log(L/A) = -2.5 \log(L) + 2.5 \log(A)$$
$$= -2.5 \log(L/r^2) + 2.5 \log(A/r^2)$$
$$= m + 2.5 \log(A/r^2)$$

for apparent magnitude m, area A, and distance r. Diameters of major and minor axes are commonly tabulated, so it is convenient to make a further modification to the equation: The area A is πR^2 for radius R or $\pi (D \times d)/4$ for apparent major and minor axes D and d, respectively. We can express the axes in arcseconds by dividing by the distance, as in the above equation. Since $2.5 \log (\pi/4) = 0.26$, the expression becomes

$$\boxed{\mu = m + 2.5 \log(D'' \times d'') + 0.26}$$

Major and minor axis galaxy diameters measured from the Palomar Observatory Sky Survey prints are tabulated directly in arcseconds in the *Uppsala General Catalogue of Galaxies* by Nilson; they may also be obtained from the RC2 and RC3 from the values for log D_{25} and the ratio of axes represented by log R_{25}. The latter gives $D'' = 10^{(\log D_{25})}/(10 \times 60)$ and $d'' = [10^{(\log R_{25})}]^{-1} \times D''$, so that

$$D'' \times d'' = \frac{[10^{\log D_{25}} \times 6]^2}{10^{\log R_{25}}}$$

For example, NGC 1300 has $\log D_{25} = 1.81$ and $\log R_{25} = 0.18$, with $B_T = 11.10$. Then $\mu = 11.10 + 2.5(4.996) + 0.26 = 23.85$ mag arcsec^{-2}.

It is often useful to convert angular measurements to linear measurements, which can be done through simple geometry if the distance is known. Consider a circle drawn with a radius equal to the distance to the galaxy, as shown in Figure 4.17. Let the galaxy diameter be represented by an arc on the circle. The ratio of the galaxy's linear diameter, d, to the circumference, $2\pi r$ for distance r, is the same as the ratio of the angular size, D, to the circumference, 360°:

$$\frac{D'}{360° \times 60' \text{ per degree}} = \frac{d(kpc)}{2\pi r(kpc)}$$

or

$$d_{kpc} = 0.29 r_{Mpc} D'.$$

We could also use the *small angle approximation*: The arc length is equal to the radius times the tangent of the angle that it subtends:

$$d \text{ (kpc)} = r \text{ (kpc)} \tan (D') \approx r(kpc) D(\text{radians})$$

A good approximation to a galaxy's distance in megaparsecs is made by dividing the observed recessional velocity of a galaxy corrected for heliocentric motions (given by the RC2 value v_o, or v_{21} in the RC3) by the Hubble constant. Tully's *Nearby Galaxies Catalog* is a useful source for velocities of nearby galaxies, because the local Virgo supercluster flow has been taken into account (details on superclusters will be discussed in Chapter 12).

In the next chapter, we will explore the distribution of light from point to point within a resolved galaxy, making use of the concepts of surface brightness and color to infer details of the distribution of stars within the galaxy.

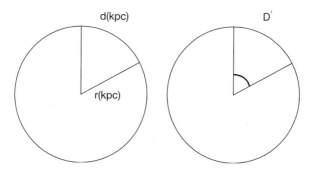

FIGURE 4.17 *The circles relate an object's linear size to its angular size.*

Exercises

1. There is a classical problem in distance determinations: Harlow Shapley, who first realized from the distribution of globular clusters that the Sun was not at the center of the Galaxy, did not know that there was dust in our galaxy. He was trying to calculate the distance to Andromeda (M31), the nearest large galaxy to ours. The apparent magnitudes of Andromeda's stars were 0.2 magnitudes fainter than they would have been in the absence of dust, which led to an overestimate of the distance to Andromeda. Determine the factor by which the distance was inaccurate because Shapley did not account for dust. (Hint: Note that the answer is not in units of distance but in percentages. Figure out the error caused by the inaccurate apparent magnitude, making use of the distance modulus.)

2. This exercise is designed to familiarize you with basic galaxy properties. Look up the galaxies (e.g., in the RC3 or via NED) in the following table, and fill in each column. (*Show your conversions on a separate sheet of paper.*)

NGC	Angular diameter (arcmin)	Distance (Mpc)	Linear diameter (kpc)	Extinction-corrected apparent blue magnitude; B_T°	Mean surface brightness μ (*mag arcsec^{-2}*)
628					
925					
1232					
1300					
2841					
3031					
4321					
5055					
5194					
5457					

3. Do a simple population synthesis model: Suppose that there is a distribution of three types of stars: O5, A0, and M0, with a Salpeter mass function. Assuming that all stars are on the main sequence, what is the (B-V) color of the region? Next, assume that the O stars have all died, and determine the resulting color. (To do this problem, use the stellar charts giving [B-V] and luminosity in V band; be sure to consider the total number of stars of each type and to convert luminosities to magnitudes.)

4. Suppose that an Sc galaxy with an inclination of 45° is observed at latitude b = 20° and has an apparent blue magnitude of 14.2 mag. Its distance modulus is 25.2, and it has a major axis diameter of 5'.
   a. Determine its corrected apparent magnitude.
   b. Determine its absolute blue magnitude.
   c. Determine its average surface brightness.

5. Determine the K correction for an E0 galaxy and an Sc galaxy with observed heliocentric velocities of 20,000 km s^{-1}.

6. Suppose an Sbc galaxy has an observed (B-V) color index of 0.9.
   a. Determine its reddening.
   b. If the visual absorption is 1.3 mag, determine the U and K band absorption.

7. Determine the linear diameter of a galaxy with an angular diameter of 5' and a distance of 20 Mpc.

Unsolved Problems

1. What determines the initial mass function?
2. How do the evolutions of different types of galaxies compare?

Useful Websites

For HST images, the Space Telescope Science Institute offers http://www.stsci.edu
A detailed list of sites for galaxy images is given in Chapter 2.
For population synthesis models by Leitherer and Heckman, email to the Space
   Telescope Science Institute and import via ftp:
   ftp.stsci.edu, login anonymous, cd observer/catalogs/pop-synthesis

Further Reading

Bershady, M., M. Hereld, R. Kron, D. Koo, J. Munn, and S. Majewski. 1994. The optical and
   near-infrared colors of galaxies. I. The photometric data. *Astronomical Journal*
   108:870.

Burstein, D., and C. Heiles. 1982. Reddenings derived from H I and galaxy counts:
   Accuracy and maps. *Astronomical Journal* 87:1165.

Charlot, S., and A.G. Bruzual. 1991. Stellar population synthesis revisited. *Astrophysical
   Journal* 367:126.

Davies, J., S. Phillipps, M. Disney, P. Boyce, and R. Evans. 1994. Understanding the surface
   brightness distribution of disc galaxies. *Monthly Notices of the Royal
   Astronomical Society* 268:984.

de Vaucouleurs, G., A. de Vaucouleurs, and H.G. Corwin. 1976. *Second reference
   catalogue of bright galaxies.* Austin: University of Texas Press.

de Vaucouleurs, G., A. de Vaucouleurs, H.G. Corwin, R. Buta, G. Paturel, and P. Fouque.
   1991. *Third reference catalogue of bright galaxies*, New York: Springer.

Elmegreen, B.G. 1985. The initial mass function and implications for cluster formation. In
   Les Houches, Session XLI: Birth and infancy of stars, Amsterdam; North Holland:
   Elsevier Science Publishers. 257.

Elmegreen, D.M. 1979. An optical analysis of dust complexes in spiral galaxies. Ph.D.
   thesis, Harvard University.

Giovanelli, R., M.P. Haynes, J.J. Salzer, G. Wegner, L.N. Da Costa, and W. Freudling. 1994.
   Extinction in Sc galaxies. *Astronomical Journal* 107:2036.

Giovanelli, R., M.P. Haynes, J.J. Salzer, G. Wegner, L.N. Da Costa, and W. Freudling. 1995.
   Dependence on luminosity of photometric properties of disk galaxies: Surface
   brightness, size, and internal extinction. *Astronomical Journal* 110:1059.

Holmberg, E. 1958. *Meddelanden från Lunds Astronomiska Observatorium*, Series 2,
   No. 136.

Leitherer, C., and T. Heckman. 1995. Synthetic properties of starburst galaxies.
   Astrophysical Journal (Suppl.) 96:9.

Leitherer, C., et al. 1996. A database for galaxy evolution modeling. *Publications of the Astronomical Society of the Pacific* 108:996.

Mihalas, D., and J. Binney. 1981. *Galactic astronomy: Structure and kinematics.* San Francisco: W.H. Freeman.

Odewahn, S., and G. de Vaucouleurs. 1992. Mean galaxy luminosity classifications. *Astrophysical Journal (Suppl.)* 83:65.

Pence, W. 1976. K-corrections for galaxies of different morphological types. *Astrophysical Journal* 203:39.

Tully, R.B. 1988. *Nearby galaxies catalog.* Cambridge: Cambridge University Press.

Sandage, A., and G. Tammann. 1981. *A revised Shapley-Ames catalog of bright galaxies.* Washington, DC: Carnegie Institution of Washington.

Scheffler, H., and H. Elsasser. 1987. *Physics of the galaxy and interstellar matter.* New York: Springer-Verlag.

van den Bergh, S. 1960. A preliminary luminosity classification for galaxies of type Sb. *Astrophysical Journal* 131:558.

van den Bergh, S. 1960. A preliminary luminosity classification of late-type galaxies. *Astrophysical Journal* 131:215.

van den Bergh, S., R. Abraham, R. Ellis, N. Tanvir, B. Santiago, and K. Glazebook. 1996. A morphological catalog of galaxies in the Hubble deep field. *Astronomical Journal* 112:359.

Vorontsov-Vel'yaminov, B.A. 1987. *Extragalactic astronomy.* New York: Harwood Academic Publishers.

Worthey, G. 1994. Comprehensive stellar population models and the disentanglement of age and metallicity effects. *Astrophysical Journal (Suppl.)* 95:107.

Wray, J.D. 1988. *The color atlas of galaxies,* Cambridge: Cambridge University Press.

CHAPTER 5
Differential Galaxy Light

Chapter Objectives: to describe measurements and interpretations of the large-scale light variation among galaxies

---

Toolbox:

radial profiles   color gradients
azimuthal profiles   arm classification
scale lengths   isophotal twists

5.1 Radial Profiles for Elliptical and S0 Galaxies

For nearby galaxies, it is possible to resolve relatively small regions and compare the flux from point to point; we refer to this as *differential light*, measured as a function of passband. Depending on our resolution, we may be able to see individual stars or only star clusters as our smallest unit. In principle, it is possible to determine the distributions and numbers of different types of stars in a galaxy through an analysis of color and flux as a function of position in the galaxy. Variations in the differential light from one type of galaxy to another and within a given galaxy reveal internal structures and differences reflecting their formation and mass distribution. Some light variations are *radial* (depending on the distance from the galaxy center), whereas others are *azimuthal* (depending on the angle in the galaxy, at a constant radius).

In ellipticals, the flux is brightest in the center, or *core*, where the most stars are concentrated, and is fainter farther out, in the *envelope*. Radial light profiles, in which intensity is plotted as a function of distance from the center to the edge, are most commonly represented by what is known as the de Vaucouleurs $r^{1/4}$ law, because the surface brightness is observed to decrease as the 1/4 power of the distance from the center. Empirical results give the light distribution as:

$$\Sigma(r) = \Sigma_e e^{-7.67[(r/r_e)^{1/4} - 1]}$$

for surface brightness Σ (in intensity units), distance r from the center, and reference values Σ_e and r_e defined below.

Converting to base 10 logarithm, the de Vaucouleurs law is given by

$$\log\left[\frac{\Sigma(r)}{\Sigma_e}\right] = -3.331\left[\left(\frac{r}{r_e}\right)^{1/4} - 1\right]$$

where r_e is the effective radius corresponding to the isophote containing half of the total luminosity, and 3.331 is a constant chosen so that this is true. With this definition, the central brightness is $10^{3.331}\Sigma_e \approx 2140\,\Sigma_e$ for surface brightness Σ_e at the radius r_e. A plot of the log of Σ/Σ_e versus $r^{1/4}$ is a straight line, as shown in Figure 5.1. The core is the region of constant surface brightness in the center.

The fit is found to match observations well for the range of radii between 0.1 r_e and 1.5 r_e, over which the surface brightness drops by about 4 mag arcsec^{-2}. This relationship is valid even for elliptical galaxies with different ellipticities, rotations, and internal velocity dispersions.

Elliptical galaxies do not follow the $r^{1/4}$ law everywhere. The cores of elliptical galaxies often have a constant surface brightness, which leads to a flat light profile there. On the other hand, the Hubble Space Telescope has revealed that

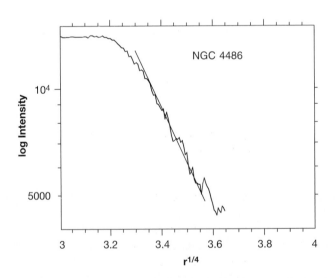

FIGURE 5.1   *Surface brightness as a function of $r^{1/4}$ for the elliptical galaxy NGC 4486, compared with an $r^{1/4}$ profile (straight line).*

the cores of some ellipticals (as well as the central regions of some spirals) have an excess of light relative to the $r^{1/4}$ law fit. Very bright ellipticals, with absolute blue magnitudes $M_B < -22$, tend to have inner surface brightness profiles that fall off with increasing r faster than $r^{1/4}$, whereas ellipticals fainter than $M_B = -22$ have slower decreases. The excess light and rapid decline are evidence for central black holes, which cause a very steep concentration of light, or *cusp*, in the accretion disk surrounding a black hole (see Chapter 11).

Some bright ellipticals also deviate from the $r^{1/4}$ law in their outer regions, with a surface brightness in excess of the expected fit. These results indicate that the structure is different for these galaxies in the inner and outer regions, although it is not well modeled or understood yet. There is evidence that mergers and capture of material and tidal disturbances from other galaxies play a role in the final structure, as we will explore more in Chapter 12.

Radial profiles in different wavelengths are all accurately represented by the de Vaucouleurs law, but the characteristic length scale may vary with wavelength. This variation shows up as a color gradient, such that E galaxies tend to be redder in the center than in the outer envelopes. (We will discuss possible reasons for this effect later.)

The universality of the radial profile for ellipticals implies that there are similar surface density distributions and underlying volume density profiles in all ellipticals. Two theoretical density distributions provide realistic fits for these profiles. The Hernquist density profile is

$$\rho(r) = \frac{\rho_0}{[r(r + r_h)^3]}$$

with $r_h = 0.55\ r_e$. The Jaffe density profile is

$$\rho(r) = \frac{\rho_0}{[r^2(r + r_j)^2]}$$

with $r_j = 1.8\ r_e$. The Jaffe profile has been applied well to observations of low mass elliptical galaxies, whereas the Hernquist model is a better fit to higher mass elliptical galaxies.

The dynamics of elliptical galaxies can be understood to first approximation by examining computer simulations of globular clusters. Globular clusters resemble elliptical galaxies except for their size; they contain only 10^5 stars instead of the 10^{11} stars in a moderate-sized elliptical galaxy and are embedded in galaxies instead of being isolated. (In some theories of galaxy formation, the galaxies are made by "assembling" many globular cluster-sized building blocks; see Chapter 12.) Ivan King has reproduced their light distribution by assuming an *isothermal* sphere of stars, in which every part of the sphere has stars moving at the same average velocity as every other part. This representation is appropriate for elliptical galaxies over their middle regions.

The bulges of S0 (lenticular) galaxies also have brightness distributions that closely fit de Vaucouleurs $r^{1/4}$ profiles, with observed profiles that can be decomposed into spherical plus flat components in the outer regions. S0 galaxies, like ellipticals, show a core (bulge) region, followed by a lens and a disk. The lens edge shows up as a bump in the radial profile, which some interpret as a ring (sometimes there may be a ring as well as a lens feature present). Farther out, a disk is evident, with an exponential rather than an $r^{1/4}$ law profile. This disk may be embedded in an outer region called an envelope, which may be a *spheroidal halo*, a *thick disk*, or an *extended disk*, depending on the details of the shape. The minor axis resembles an elliptical galaxy, with an $r^{1/4}$ light profile, whereas the major axis resembles a disk system, with an exponential profile in the outer parts.

Statistically, there should be equal numbers of lenticular galaxies at all ellipticities. Recent studies indicate that apparently round lenticular galaxies are proportionately less common than flattened ones. This discrepancy suggests that there may be some confusion in classifying face-on lenticulars. In S0 galaxies, the disks are difficult to identify by eye alone; some may be incorrectly classified as E0–E3 galaxies.

5.2 Azimuthal Profiles for Elliptical Galaxies

Measurements of the *azimuthal* light distribution in a galaxy, which is the variation of light at a given radius as a function of angle around the galaxy, provide insight into the way a galaxy forms or evolves. For example, two-dimensional symmetry implies a dominance of rotational motions, whereas unequal axes imply that random motions (anisotropic velocity dispersions) rather than rotation support the galaxy; galaxy kinematics will be discussed in Chapter 7. Sometimes two axes are nearly the same and longer than a third, which makes the galaxy oblate like a saucer, and sometimes two axes are nearly the same and much shorter than the third, which makes them prolate like a cigar. In some galaxies no two axes are the same length, so they are triaxial.

Ellipticals with smooth brightness distributions have apparent isophotes, which are contours of equal apparent brightness, in a series of nested ellipses, as shown in Figure 5.2. An axially symmetric elliptical galaxy shows a symmetric radial profile no matter what the orientation of the radial cut, and it has symmetric contours.

In recent years, two prominent morphological subdivisions of ellipticals have emerged: *boxy* and *disky* ellipticals. The differences in their morphology show up as deviations in their isophotal contours, or in the radial profiles made along different angles in the galaxy. Sketches of the two types are shown in Figure 5.3. Boxy ellipticals have light distributions that trace out a rectangular region in the isophotes, whereas disky ellipticals have a distinctly pointed component in the outer isophotes.

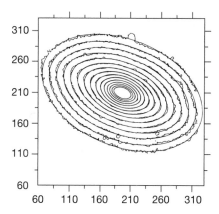

FIGURE 5.2   *Contours of the elliptical galaxy NGC 4697, with smooth ellipses. (From models, Carter 1987.)*

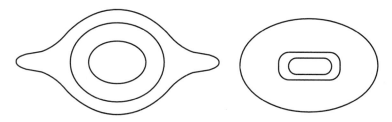

FIGURE 5.3   *Schematic diagram of disky (left) and boxy (right) ellipticals.*

The intensity of ellipticals can be expressed as

$$I = I_o + A_n\sin(n\theta) + B_n\cos(n\theta)$$

where θ measures the position angle of the major axis. The A and B coefficients for $n > 1$ represent the amplitudes of the deviations from perfect ellipticity, which are typically around 1%. The parameters A_4 and B_4, which correspond to the coefficients of the fourth-order harmonic terms, indicate whether a galaxy is boxy (negative amplitude deviations) or disky (positive amplitude deviations).

Data indicate that boxy ellipticals are slowly rotating and have large anisotropic velocity dispersions; they may be interacting galaxies. In contrast, disky ellipticals are rotating faster and may be more isolated; they are an extension of the S0 classification. In Chapter 7 we will explore the so-called *fundamental plane* of ellipticals, which relates the velocity dispersion to the surface brightness and radius.

5.3 Isophotal Twists in Elliptical Galaxies

The ellipticity of isophotes may vary with radius in the inner regions. Also, there may be a variation of major axis position angle (PA) of these isophotes with radius. The PA gives the orientation of the galaxy in the sky; it is the angle measured counterclockwise from north to the major axis, as shown in Figure 5.4.

A variation of PA with radius, or a twist, may be an indication of triaxiality, with no axis of rotational symmetry; see Figure 5.5. The bulge in M31 shows a twist in the major axis by about 10° near its center and has been fit by a triaxial model. M81 shows a similar small central twist, and near-infrared observations indicate that this may be common in spirals. Such twists may also be in-

FIGURE 5.4 *Position angle is measured from the north counterclockwise to the major axis of a galaxy.*

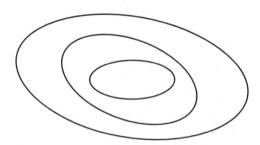

FIGURE 5.5 *Some galaxies show isophotal twists and ellipticity variations with radius.*

terpreted as an indication of the presence of small bars or nonaxisymmetric bulges in the central regions. This subject will be considered more fully in Section 5.10.

5.4 Radial Profiles for Spiral Galaxies

Spiral galaxies have bulges that are very similar to elliptical galaxies. The bulge surface brightness generally has the form $\log (\Sigma/\Sigma_o) \propto r^{-1/4}$, the same as for elliptical galaxies, although recent studies suggest that an exponential distribution may be at least as good a fit to the central light. Sometimes the bulge region has a boxy shape, just as some ellipticals do, as shown in Figure 5.6.

Almost every disk galaxy with a measured radial profile, including Magellanic irregulars, shows an exponential profile in the disk. In the region beyond the bulge, the brightness decreases approximately exponentially as:

$$\Sigma = \Sigma_o e^{(-r/r_s)}$$

where Σ_o is the intercept value, equal to the extrapolated central surface brightness, and r_s is the *scale length*. This is sometimes written in the form $\Sigma = \Sigma_o e^{-\alpha r}$, where $\alpha = 1/r_s$. Figure 5.7 shows a radial profile for a spiral galaxy.

A scale length is the distance over which light decreases by $1/e$, or by about a factor of one-third. A galaxy with a larger scale length has a slower decrease in brightness than a galaxy with a smaller scale length. In Figure 5.7, the light decreases from 1000 (in arbitrary units) in the center to $(1000/e) = 368$ at a radius of about $0.2\ R_{25}$. (It is customary to measure the scale length in terms of R_{25}, so that you do not need to know the distance to the galaxy.) Typically $r_s/R_{25} \approx 0.2$–0.3, so there are 3–5 scale lengths out to R_{25}. That is, spiral arms can usually be traced out to a distance of ~4 scale lengths in a galaxy on typical prints. Recent studies indicate that blue scale lengths may sometimes be shorter than near-

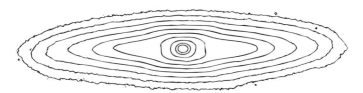

FIGURE 5.6 *Magnitude contours are shown for the edge-on spiral galaxy NGC 2310, which has a boxy bulge. (From Shaw 1993.)*

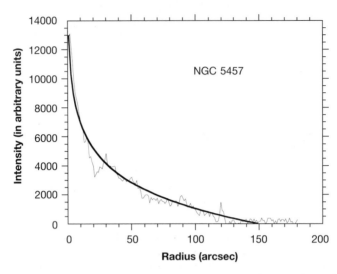

FIGURE 5.7 *Radial profile for the spiral galaxy NGC 5457, with a solid curve showing an exponential light profile for comparison.*

infrared scale lengths. This fact has consequences for inferred characteristics of the stellar populations as well as dust distributions.

A logarithmic plot of surface brightness versus radius is a straight line whose slope is a measure of the inverse scale length, as shown in Figure 5.8.

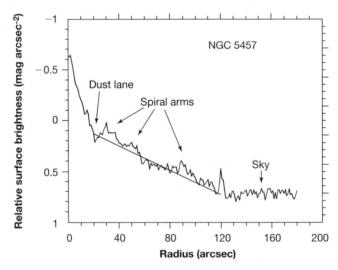

FIGURE 5.8 *Logarithmic plot of the radial profile of NGC 5457, compared with the straight line from theory.*

In terms of surface brightness, the scale length is given by

$$\mu(r) = \mu_o + 1.086 \frac{r}{r_s}$$

for central surface brightness μ_o. This expression results from the use of log (base 10) in defining magnitudes, and ln (base e) in defining scale length: the scale length r_s is given by $r/r_s = \ln \Sigma/\Sigma_o = 2.3 \log \Sigma/\Sigma_o$. In Figure 5.8, the inverse slope is 1.0/[2.3 (3–1)], so the scale length is ~0.2, in agreement with the value found from Figure 5.7. The central regions of a galaxy are a combination of an $r^{1/4}$ shape plus an exponential disk. In deep exposures, these regions may be saturated, so the central extrapolated surface intensity may be difficult to determine. With a log plot, the outer slope determines the scale length without needing to know the central surface brightness, so it is easier to use.

The nearby spiral galaxies that are most familiar to us all have the same central blue surface brightness, 21.65±0.3 mag arcsec^{-2}; this result is known as *Freeman's law*. Recent results indicate that this is an upper limit rather than an absolute range for spirals, and some have suggested that it is no more than a selection effect. If the central disk surface brightness in normal spirals is approximately constant, the total absolute magnitude of galactic disks is determined by the scale length. Low surface brightness (LSB) galaxies, on the other hand, have typical blue central surface brightnesses of 23.8 mag arcsec^{-2}.

Ken Freeman noted that the transition region between the bulge and disk may have either of two forms: Type I has an excess of light just outside the bulge relative to the extrapolated exponential disk; this excess may be due to the presence of a bar or ring. Type II has an apparent dip in the inner disk (also seen in CO in some galaxies; see Chapter 6), which could result from orbit perturbations near a bar or from mass infall to the nucleus. Good empirical fits to Type II profiles can be achieved by combining an $r^{1/4}$ profile with an exponential disk of the form

$$\Sigma = \Sigma_o \exp\left[- \alpha r + \left(\frac{\beta}{r}\right)^3 \right]$$

which results in a low disk luminosity in the regions interior to β.

Typical scale lengths for normal spiral galaxies are $r_s = 2$–5 kpc, independent of Hubble type, and $r_e \sim 0.5$–4 kpc, depending on Hubble type (r_e is bigger in earlier Hubble types); in all cases, $r_e < r_s$. This relation is observed for LSB galaxies as well. For the Milky Way, the current best estimates for the scale lengths are 5.0±0.5 kpc for r_s and 2.7 kpc for r_e.

TABLE 5.1   *Disk/bulge ratios*

Hubble type	L_{bulge}/L_{total}	D/B	Hubble type	L_{bulge}/L_{total}	D/B
Sb	0.45	2.2	Sc	0.15	6.7
Sbc	0.32	3.1	Sd	0.01	100

Source: From Kent 1987 and Boroson 1981.

Typically, more than 75% of the total blue light of a spiral galaxy comes from the disk. An evaluation of the ratio of disk to bulge luminosity (D/B) through an integration of the radial light distributions yields

$$\frac{D}{B} = 0.28 \left(\frac{r_s}{r_e}\right)^2 \frac{\Sigma_0}{\Sigma_e}$$

(See Mihalas and Binney 1981 for further details.)

The typical D/B ratio is about 1 for S0 galaxies, but with wide variation; the ratio for Sb galaxies (e.g., M31) is about 3, and for Sc galaxies (e.g., M33) about 12. Table 5.1 gives representative values for disk/bulge ratios, along with bulge-to-total luminosity ratios, based on a decomposition of the radial profiles into the bulge and disk components. For comparison, the Milky Way has a bulge luminosity/total luminosity ratio of 0.34±0.08, which implies type Sbc. LSB galaxies generally have D/B > 10, although a D/B ~ 1 is not uncommon.

The perpendicular light profiles of disk galaxies have been measured in edge-on systems. They show an approximately exponential light decrease with increasing z distance that is independent of radial distance from the galactic center in the disk. The disks of some galaxies, particularly S0s, have an extensive flat component referred to as a *thick disk*.

5.5 Radial Profiles for Barred Spiral Galaxies

Barred galaxies, like nonbarred spirals, have exponential disks. There are two types of bars: *Flat bars* have almost constant surface brightness along the bar; that is, they have a much shallower decline than the disk. An example is NGC 1300 shown in Figure 5.9. *Exponential bars*, in contrast, have the same scale length as the disk; an example is NGC 2500 in Figure 4.9.

Figure 5.10 shows the general shape for bar radial profiles. The solid curve indicates the azimuthally-averaged exponential profile for the disk. The profile along the bar is shown by the top line; the bar ends at a radius of about 0.2 R_{25} in this figure. Beyond the bar, the line represents the intensity along the spiral arms, which is higher than that along the interarm regions. The light distribution along the exponential bar and the continuation into the spiral arms are shown in the intermediate curve; the bar does not stand out in this case. For compari-

FIGURE 5.9 *Flat-barred galaxy NGC 1300. (Imaged in the B band with the KPNO Burrell-Schmidt telescope. From Elmegreen et al. 1995.)*

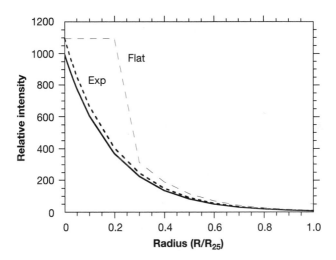

FIGURE 5.10 *Radial profiles in spiral galaxies with strong (flat) and weak (exponential) bars.*

son, radial profiles are shown for the flat-barred galaxy NGC 1300 and the exponential-barred galaxy NGC 1359 in Figure 5.11.

A number of galaxy parameters are correlated with bar type. There is a strong Hubble-type dependence, in the sense that early types usually have flat bars. A strong Arm Class dependence (see Section 5.8) is also present: flat-barred galaxies usually have a grand design spiral structure, whereas exponential-barred galaxies usually are flocculent in their outer disks. Furthermore, flat bars gener-

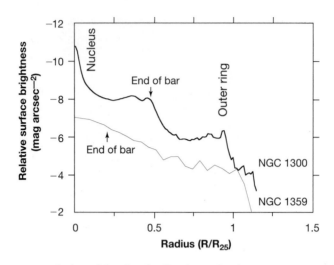

FIGURE 5.11 *Radial profiles for the flat-barred galaxy NGC 1300 and the exponential-barred galaxy NGC 1359.*

ally are larger compared with the overall galaxy size than exponential bars: Flat bars have $r_{bar}/R_{25} > 0.4$, whereas exponential bars have sizes $r_{bar}/R_{25} < 0.3$. All of these factors are consistent with the idea that big flat bars drive the spiral patterns, which will be considered in more detail in Chapter 9.

Recent infrared studies of some galaxies reveal tiny bars that are not detectable optically, as shown in the K band image of NGC 6946 in Figure 5.12.

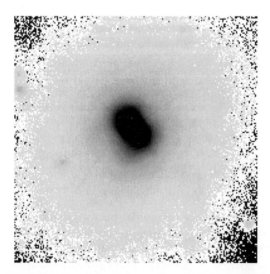

FIGURE 5.12 *This K band image of the center of the spiral galaxy NGC 6946 reveals a tiny stellar bar inside the circular region. (From the KPNO 2.1-m telescope, D. Elmegreen and F. Chromey, Vassar College.)*

These observations suggest that tiny bars contain very old stars, sometimes obscured by dust at optical wavelengths. Maps of CO distributions often reveal tiny bars in the gas also.

5.6 Image Enhancement

Galaxies are randomly inclined to our line of sight, and high inclinations make it difficult to discern spiral structure. We can simulate face-on aspects by "stretching" images until the galaxies appear round (since disks presumably are circular). Computers facilitate the rectification of images. If the image is stored as an array of numbers (which are picture elements, or *pixels*), deprojection is accomplished by multiplying the pixels along the minor axis by a factor equal to the inverse of the cosine of the inclination, i. The corrected distance along the minor axis, d_c, is then given in terms of its observed distance d_{obs} by

$$d_c = \frac{d_{obs}}{\cos(i)}$$

Stretching the minor axis on computer is easiest if the image is oriented so that its minor axis is either horizontal or vertical. This orientation is achieved by rotating the galaxy image through an angle equal to its position angle. The major axis, of course, is undistorted. The resulting deprojected image of a spiral galaxy should be circular, in general, although sometimes galaxies have rings that are not circular.

Photographic reproductions have a limited dynamic range, so images usually end up overexposed in the center or too faint in the outer parts. It is important to note that spiral arms appear very weakly on radial plots: On an averaged radial profile (that is, an average over all azimuthal angles) they would hardly be noticeable, and even a direct cut would show only a small blip for most galaxies. The point is that spirals are bright only relative to their immediate neighborhood, typically by about a magnitude or so. The images can be enhanced by removing the azimuthally averaged radial profile from each galaxy and normalizing each radial position to a constant average deviation so that features in both the inner and the outer regions can be seen equally well. This enhancement is done on a digital galaxy image by dividing the intensity of the light at each pixel by the average light at that radius from the galaxy center. The resulting light is normalized at each radius to a constant rms value.

After enhancements, remarkable structure becomes apparent that was hidden in the original images. In the deprojected, radial profile–subtracted image of M81, shown in Figure 5.13, note the narrowing of the two main arms halfway out, where the arm brightness decreases. (Compare this image with the conventional sky view of M81 in Figure 5.16.) This pattern has been noticed in

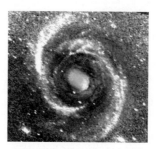

FIGURE 5.13 *An enhanced print of M81 shows brightness variation along the arms. Compare this with Figure 5.16 (left), which is a negative image of the same galaxy without enhancement. (From Elmegreen et al. 1992.)*

several galaxies, and it turns out to be an important indicator of internal motions in a galaxy. Further discussion of the brightness variation along spiral arms is in Chapter 9, Section 9.4.

5.7 Logarithmic Spirals

Galaxies have approximately logarithmic arms, which means that the pitch angle of the arms is roughly constant regardless of where it is measured. Another effective computer technique for analyzing arm structure is to plot the galaxy image in an "unwrapped" form, either as a polar plot (converting x, y coordinates to r, θ), or in a (log r, θ) plot. These types of plots straighten out the arms and allow their degree of symmetry and continuity to be visually assessed more easily.

Figure 5.14 shows a (log r, θ) plot of NGC 628, which is a multiple-arm galaxy with long arms. Their regularity can be seen in the evenness of their spacing. The plot extends from 0° to 720° for clarity.

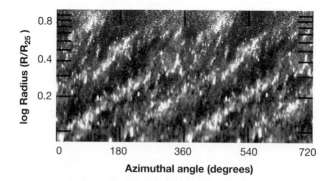

FIGURE 5.14 *NGC 628 is shown unwrapped, with log (radius) increasing upward (tic marks represent 0.1 R_{25}) and azimuthal angle θ increasing to the right. Compare this with Figure 9.13, which is a sky view of NGC 628.*

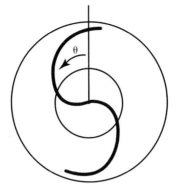

FIGURE 5.15   *Arm angles.*

The slope of the arms on these plots is equal to tan i for pitch angle i (defined in Section 2.4). This characteristic shape was predicted by models of spiral density waves by W. Roberts, M. Roberts, and F. Shu:

$$r = r_0 \exp(\theta \tan i)$$

where θ (in radians here) is the azimuthal angle at which an arm appears at a given radius. In Figure 5.15, the angle θ is measured counterclockwise from north to the arm, at each radius.

Some theories produce spirals whose pitch angle varies with radius; for example, the theory of stochastic self-propagating star formation (described in Chapter 9) yields a hyperbolic function in which the pitch angle i increases with radius. In fact, in this model, the angle times the radius, θr, remains constant, so $d\theta/d(\log r) \propto 1/r$. In practice, hyperbolic spirals are so similar to logarithmic spirals (to within measuring errors) that it is hard to differentiate between the two models. Observations suggest that the typical spacing between arms is ~ 1–2 kpc.

5.8 Arm Classification

A classification scheme developed by Debra and Bruce Elmegreen addresses details of spiral arm structure. This *Arm Classification* system assesses the degree of arm symmetry and continuity. Galaxies with arms that are long, continuous, and symmetric are called *grand design*; those with numerous short and asymmetric arms are called *flocculent*, because of their fleecy appearance. Intermediate types showing characteristics of both flocculent and grand design galaxies are called *multiple-arm* galaxies. Multiple-arm galaxies have inner two-arm symmetries, but the outer regions are highly branched and include many spiral arms and arm segments. Figure 5.16 shows schematic diagrams of the main classes, and three galaxies as examples.

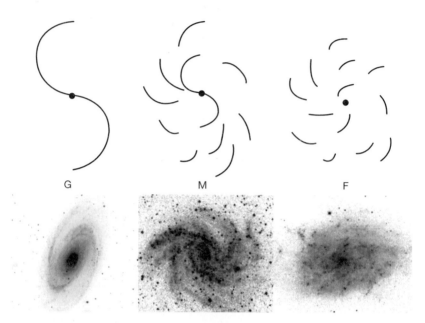

FIGURE 5.16   *(left) Grand design galaxy, M81; (middle) Multiple-arm galaxy, NGC 6946; (right) Flocculent galaxy, NGC 7793. (Photos taken with the Palomar 1.2-m Schmidt telescope by the author.)*

A galaxy's arm classification is essentially independent of its Hubble type; both early and late Hubble types can be either flocculent or grand design. However, there is a preponderance of flocculent galaxies among later Hubble types. The Arm Classification system is also independent of the Luminosity Class, but there is some overlap in the traits at the low mass end because small, low luminosity, nearly irregular spiral galaxies tend to have flocculent structure. The classification system further subdivides galaxies into one of 12 types, with various structural distinctions that may or may not be physically important.

Surface photometry measurements, which reveal the light variations across the disk, suggest that there is a physical significance to the arm classification system: grand design galaxies are inferred to contain a *spiral density wave*, which is a global gravitational perturbation that causes long symmetric spiral arms to develop as material is temporarily collected or organized along the spiral perturbation wave crests. Multiple-arm galaxies have a coherent two-arm pattern in the inner regions, with some complications in the outer regions. Flocculent galaxies probably lack a global spiral density wave and may acquire their shape through stochastic or random star formation processes, where a short sequence of star-forming regions is drawn out into an arc because of differential rotation. The physics underlying these morphologies will be discussed further in Chapter 9.

Recent K-band images have revealed a weak two-armed pattern in some flocculent galaxies, which suggests that their underlying structure is density wave–driven but that recent random star formation or many overlapping modes have led to a chaotic young spiral structure. There may be physically different types of flocculent galaxies.

5.9 Azimuthal Profiles in Spiral Galaxies

It is difficult to discern spiral structure on radial profiles because the exponential light fall-off dominates. On azimuthal profiles, however, spiral arms are clearly brighter than the adjacent interarm regions at that radius. If the arms are symmetric, then a plot of intensity versus angle around the galaxy displays peaks that are evenly separated: two main arms are separated by 180°, three main arms are separated by 120°, and so on. It is most common for there to be two symmetric arms (see Figure 5.17). For a wide range of galaxy types, the inner regions show two approximately symmetric arms, which then bifurcate midway out in the disk. (There may be other arms or spurs in the inner regions as well, but in multiple-arm or grand design galaxies two arms are most prominent.)

In a galaxy that has a grand design, observations show that the arms are enhanced nearly as much in the red as in the blue. Because red light comes mostly from the older (lower mass) disk stars, and blue light comes mostly from the younger (higher mass) stars, nearly equal intensity contrasts in the red and blue indicate the presence of density waves that organize almost all of the material in the disk, both young and older stars, into coherent patterns. Azimuthal profiles

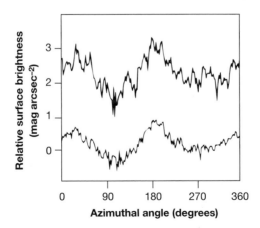

FIGURE 5.17   *B band (upper line) and I band (lower line) azimuthal profiles for the grand design galaxy NGC 4321, with arbitrary zero point for the surface brightness. Note the separation of the main arms by 180°.*

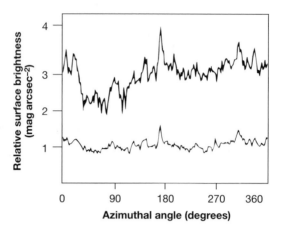

FIGURE 5.18 *B band (upper line) and I band (lower line) azimuthal profiles for the flocculent galaxy NGC 5055, with arbitrary zero point for the surface brightness. Note the lack of regularity and the higher-amplitude B peaks. (From Elmegreen and Elmegreen 1984.)*

of flocculent galaxies are easily distinguished from those of grand design galaxies: In flocculent galaxies, the arm segments are very blue and irregularly spaced (see Figure 5.18). The underlying red disk of older stars is much more homogeneous, so the arm-interarm contrast is smaller in red light than in blue light.

Although the disk brightness decreases exponentially, the arms generally do not decrease in brightness quite as fast as the interarm regions. Because spiral arms are rather blue owing to the presence of the high mass stars found preferentially in them, the arms become even brighter than the interarm regions with increasing radius in nonbarred spiral galaxies, and so the sum of the arm plus disk light is slightly bluer farther out than closer in.

5.10　Arm-Interarm Contrasts

The arm-interarm contrast, Δm, measures how much brighter the arm is than an adjacent interarm region. In general, the arm-interarm contrast is strongest between the inner region and corotation (where stars and gas are moving at the same rate as the density wave pattern), and is weaker farther out. The near-infrared contrast, Δm_I, gives an indication of the strength of the spiral density wave, on the basis of matches of observations with calculations of expected contrasts from compression of disk material. The physics underlying density waves will be covered in Chapter 9 after disk kinematics has been discussed; in brief, these waves lead to density enhancements at the wave crests, which move through the disk. If a density wave has organized the material in a disk, then both young and old stars will show evidence of the perturbation; that is, both blue and

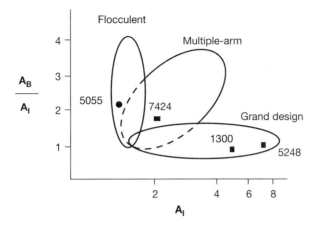

FIGURE 5.19   *Arm-interarm contrasts in the B and I bands give an indication of the presence and strength of spiral density waves. Maximum contrasts measured for several galaxies are indicated. $A_B = 10^{0.4\Delta m_B}$, $A_I = 10^{0.4\Delta m_I}$. (Adapted from Elmegreen and Elmegreen 1984.)*

near-infrared light will show similar enhancements. On the other hand, if a density wave is weak or absent, the near-infrared arms (representing the older disk stars) have a smaller contrast.

Observations show that flocculent galaxies have high Δm_B, low Δm_I, which implies that the prominent arm features are due to star formation rather than a disk density enhancement. Grand design galaxies have similar arm-interarm contrasts in both the blue and the near-infrared, because all stars are compressed by a density wave. Furthermore, the contrast can be very large for a strong density wave. Multiple-arm galaxies show characteristics intermediate between flocculent and grand design. A plot of the ratio of the blue to near-infrared arm-interarm intensity versus the near-infrared arm-interarm intensity, known as a *color-amplitude plot*, shows quantitative differences between arm classes. A schematic diagram is shown in Figure 5.19.

In barred spiral galaxies, the bar-interbar contrast increases with radius for both exponential and flat bars. Bar surface brightnesses, like arm surface brightnesses, may be as much as a magnitude brighter than the adjacent light.

5.11 Isophotal Twists in Spiral Galaxies

Isophotal twists occur in the central regions of many barred spiral galaxies. We saw in Section 5.3 that such twists may indicate triaxial or minibar structure in the innermost regions of a bulge. In a bar, twists as great as 90° are observed in many early-type galaxies. Such twists are observed in optical and near-infrared passbands, as shown in Figures 5.20 and 5.21. The fact that no late-type galaxy

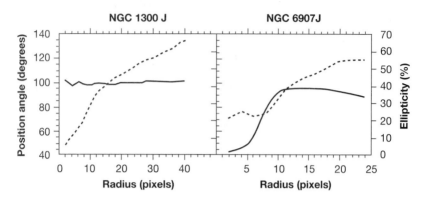

FIGURE 5.20   *NGC 1300 and NGC 6907 are both flat-barred galaxies. Their position angles are shown as solid lines, and their ellipticities as dashed lines. NGC 6907 shows an isophotal twist, indicated by the solid line jump at a radius of 9 pixels, whereas NGC 1300 has a constant position angle throughout. Both have ellipticities that increase with radius, from the round central region to the elongation resulting from the bar. (From Elmegreen et al. 1996.)*

has yet been observed with a twist may be a clue to the origin of the twist. Some computer studies indicate that twists can occur near an inner Lindblad resonance, if one is present. This resonance will be discussed in Chapter 9; it is generally associated with a large bulge, so late-type galaxies, which have small bulges, may lack inner resonances. The twists may be due to bars within bars, such that a barred galaxy has embedded in the bar another smaller, offset bar rotating at a different rate. Twisting can also occur if gas is trapped near a resonance. The collection of gas can lead to more star formation there. In some galaxies, dust may cause an apparent twist.

In Chapter 6 we will explore details of the gaseous components of galaxies.

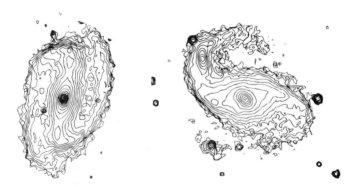

FIGURE 5.21   *J band contour plots: (left) NGC 1300 (no isophotal twist), (right) NGC 6907 (with isophotal twist). (From Elmegreen et al. 1996.)*

Exercises

Note: Many of these exercises mention the use of IRAF, the Image Reduction and Analysis Facility used at national observatories. It can be imported from http://www.noao.edu.These exercises may also be completed using other image processing software, such as IDL.

1. Verify the $r^{1/4}$ "law" for ellipticals: Plot the surface brightness as a function of radius$^{1/4}$ for the isophotal contours of an elliptical galaxy. You may use the one in Figure 5.2, or in Figure 2 of M. Liller (1960, *Astrophysical Journal* 132:309), or import your own elliptical via the Digital Sky Survey website (http://www.stsci.edu). Measure the radius along the major axis for each isophote. Convert to arcseconds using the scale line shown. Plot log I vs. $r^{1/4}$ to see that it is a straight line.

2. Import an image of NGC 3031 from the Space Telescope Science Institute public archives (http://www.stsci.edu) to your computer. Deproject it. In IRAF you will have to ROTATE it, so that its major axis is either vertical or horizontal, and then MAGNIFY it by stretching the minor axis by the ratio of major to minor axes, as listed in RC3. Print out the deprojected image.

3. Make a radial profile through the center of the galaxy in Exercise 2 using PVECTOR. Print it out.

4. Determine the disk scale length from your radial profile in Exercise 3. Make sure the sky background has been subtracted. If it has not, subtract an average sky value (or you can subtract the sky value from each of your measured points after you have made the profile); in IRAF, you can do this subtraction on the image using IMARITH. Determine the exponential disk scale length, measured as the distance where the light has decreased by 1/e. Do this by measuring just the left-hand side (more accurate would be to do both sides and average, of course, but that is not necessary here). Draw a smooth line connecting the bottom of the curve (i.e., ignore the spiral arm spikes indicated). Measure the intensity at several radial positions. Plot I vs. r, where r is the distance from the center of the galaxy. Guess what the central extrapolated value I_0 is, then find the scale length r_s (where $I = I_0/e$). Look up R_{25} for this galaxy (in RC3 or NED, for example) and convert your pixels to fractions of R_{25}. What is r_s in terms of R_{25}?
   Now plot (ln I) vs. r. Measure the slope a; express it as a positive number. The scale length r_s is the inverse of the slope. What is r_s in terms of R_{25} when you measure it this way? (This is more accurate but should be the same as the first way you did it.)

5. From the azimuthal plots of NGC 4321 and NGC 5055 in Figures 5.17 and 5.18, calculate the arm-interarm contrast in B and I. (Draw a baseline for the interarm region, through the bottom of each profile; subtract this from the arm peaks that you estimate; do this for each arm.)

6. Make a color-amplitude plot of the results from Exercise 5 by dividing the contrast in the blue by the contrast in the infrared and plotting versus I band results.

7. The IRAF task ELLIPSE measures the ellipticity and position angle of galaxies by Fourier analysis of the isophotes. Import a spiral galaxy image via the Sky Survey archives and determine the ellipticity and position angle as a function of radius. Repeat the procedure until the errors in the coefficients are less than 0.1.

8. Use the IRAF task ELLIPSE to determine the average galaxy intensity as a function of radius for some image that you have obtained or imported via the web. Construct a smoothed galaxy by making this radial fit into a two-dimensional image via the IRAF task MKIMAGE. Divide the original galaxy image by this image to normalize the light as a function of radius, in order to highlight faint features.

Unsolved Problems

1. What are the three-dimensional structure and extent of different types of galaxies?
2. What is the physical origin of the de Vaucouleurs law and exponential profiles?

Useful Websites

See Useful Websites in Chapter 2

Also, R. de Jong's bulge/disk decompositions are available via:

http://astro.u-strsbg.fr/cgi-bin/Cat?J/A+AS/118/557/

and his luminosity profiles are available at:

http://astro.u-strasbg.fr/cgi-bin/Cat?J/A+AS/106/451/

Further Reading

Boroson, T. 1981. The distribution of luminosity in spiral galaxies. *Astrophysical Journal (Suppl.)* 46:177.

Burkert, A. 1993. Do elliptical galaxies have $r^{1/4}$ brightness profiles? *Astronomy and Astrophysics* 278:23.

Carter, D. 1987. Weak disks in rapidly rotating elliptical galaxies. *Astrophysical Journal* 312:514.

de Jong, R.S. 1996. Near-infrared and optical broadband surface photometry of 86 face-on disk dominated galaxies. II. A two-dimensional method to determine bulge and disk parameters. *Astronomy and Astrophysics (Suppl.)* 118:557.

Elmegreen, B.G., D.M. Elmegreen, and L. Montenegro. 1992. Optical tracers of spiral wave resonances in galaxies. II—Hidden three-arm spirals in a sample of 18 galaxies. *Astrophysical Journal (Suppl.)* 79:37.

Elmegreen, D.M., and B.G. Elmegreen. 1984. Blue and near-infrared surface photometry of spiral structure in 34 nonbarred grand design and flocculent galaxies. *Astrophysical Journal (Suppl.)* 54:127.

Elmegreen, D.M., and B.G. Elmegreen. 1987 Arm classifications for spiral galaxies. *Astrophysical Journal* 320:183.

Elmegreen, D.M., B.G. Elmegreen, F. Chromey, D. Hasselbacher. 1996. Star formation in the outer resonance ring of NGC 1300. *Astrophysical Journal* 469:131.

Elmegreen, D.M., B.G. Elmegreen, F. Chromey, D. Hasselbacher, and B. Bissell. 1996. Near-infrared observations of isophotal twists in barred spiral galaxies. *Astronomical Journal* 111:1880.

Freeman, K. 1970. On the disks of spiral and S0 galaxies. *Astrophysical Journal* 160:811.

Gilmore, G., I. King, and P. van der Kruit. 1989. *The Milky Way as a galaxy: 19th Saas-Fee course.* Geneva: Geneva Observatory.

Graham, A., T. Lauer, M. Colless, and M. Postman. 1996. Brightest cluster galaxy shapes. *Astrophysical Journal* 465:534.

Heraudeau, P., and F. Simien. 1996. Optical and I-band surface photometry of spiral galaxies. I. The data. *Astronomy and Astrophysics (Suppl.)* 118:111.

Kent, S. 1987. Surface photometry of six local group galaxies. *Astrophysical Journal* 94:306.

Kormendy, J., and S. Djorgovski. 1989. Surface photometry and the structure of elliptical galaxies. *Annual Review of Astronomy and Astrophysics* 27:235.

Liller, M. 1960. The distribution of intensity in elliptical galaxies of the Virgo cluster. *Astrophysical Journal* 132:309.

McGaugh, S., J. Schombert, and G. Bothun. 1995. The morphology of low surface brightness disk galaxies. *Astronomical Journal* 109:2019.

Michard, R., and J. Marchal. 1993. Quantitative morphology of E–S0 galaxies. I. Bulge, lens, disk, and envelope in edge-on systems. *Astronomy and Astrophysics (Suppl.)* 98:29.

Mihalas, D., and J. Binney. 1981. *Galactic astronomy: Structure and kinematics.* San Francisco: W.H. Freeman.

Rix, H.-W. 1993. Mapping the stellar backbones of spiral galaxies. *Publications of the Astronomical Society of the Pacific* 105:999.

Schweizer, F. 1976. Photometric studies of spiral structure. I. The disks and arms of six Sb I and Sc I galaxies. *Astrophysical Journal (Suppl.)* 31:313.

Shaw, M. 1993. The photometric properties of "box/peanut" galactic bulges. *Monthly Notices of the Royal Astronomical Society* 261:718.

Zaritsky, D., and K. Lo. 1986. Evidence for nonaxisymmetric nuclear bulges in spiral galaxies. *Astrophysical Journal* 303:66.

CHAPTER 6

Gas

Chapter Objectives: to describe the gas phases and distributions in different types of galaxies

---

Toolbox:

phases of the ISM
column density

hyperfine transitions
metallicity

6.1 Radiation from Neutral Atomic Gas

The *interstellar medium* (ISM) is the material between the stars in a galaxy. It consists of intermixed gas and dust, with about 100 times more gas than dust by mass. The *phases* of the interstellar medium include a cold component, a warm component, and a hot component; the gas may be in the form of atoms, molecules, or ions. Some gas particles are diffusely distributed throughout the disk; others are clumped into big or small clouds. The average disk density is only about 1 atom per cubic centimeter.

Hydrogen atoms are the primary form of ordinary matter. They radiate strongly in the centimeter (radio) part of the electromagnetic spectrum, owing to electron jumps between split levels in the ground state. This split, or *hyperfine structure*, is caused by the slightly different energies an electron has depending on whether its spin is in the same direction as the proton spin or in the opposite direction, as shown in Figure 6.1. Slightly more energy is required for the electron to have the same spin (called the *parallel state*) than the opposite spin (*antiparallel state*). In the antiparallel state, the electron and proton, being oppositely charged, behave like the north and south pole of two magnets brought near each other. The parallel state is analogous to two north poles brought together; there is a slight repulsion, so this is a higher energy state.

A hydrogen atom that collides with another particle (or acquires sufficiently energetic radiation) can undergo a *spin-flip transition*, in which the electron flips its spin orientation and moves to the higher energy level of the ground state. The electron is very unlikely to flip back spontaneously to its original spin

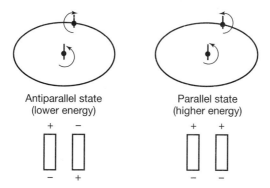

Antiparallel state
(lower energy)

Parallel state
(higher energy)

FIGURE 6.1 *Hyperfine transitions in a hydrogen atom.*

orientation; on average in the interstellar medium, a spontaneous decay for any given atom occurs about once every 10^7 years. That same atom will be hit about every million years on average, and then the colliding particle will cause deexcitation and carry the extra energy away. Because there are so many hydrogen atoms, however, many are able to spontaneously decay, and when they do, they emit a photon with a wavelength $\lambda = 21$ cm (which corresponds to a frequency $\nu = c/\lambda = 1420$ Mhz).

Atomic clouds, called *H I clouds* because they are mostly neutral hydrogen, have temperatures of 80 K–100 K as a result of a balance between heating from incoming stellar radiation and cooling from cloud radiation. The dominant form of cooling comes from radiation by collisionally excited carbon atoms that are intermixed with the hydrogen atoms. In addition to atomic clouds, there is a pervasive distribution of warm intercloud neutral gas with temperatures of about 1000 K.

Gas emits all along a given line of sight, so that the total thickness or geometry of a radiating region is not always determinable. Instead of speaking of the number density, n (number per cubic centimeter), of material in a given direction, it is customary to speak of the *column density*, N (number per square centimeter), which is the material integrated along the line of sight of length L:

$$N(\text{cm}^{-2}) = \int n \, dL$$

The column density is measured in terms of the observed brightness of a region, which depends on the cloud temperature and the background source (if any) temperature, moderated by the optical depth at the wavelength of interest. The 21-cm emission is received over a range of frequencies (or velocities) because relative motions of material in the disk lead to Doppler-shifted emission.

If the atomic hydrogen line is not saturated (that is, if the gas is optically thin, as atomic hydrogen usually is), then the column density $N(\text{H I}) \propto T \, \Delta v$, where Δv

is the *full width at half maximum* (FWHM) of the line. More specifically, the column density is given by

$$N(H\ I\ cm^{-2}) = 1.823 \times 10^{18} \int T_B(H\ I)\ dv$$

for *brightness temperature* T_B (K) and linewidth dv (km s^{-1}), which integrates the area under the emission line curve, as shown in Figure 6.2. Saturation means there are so many atoms along the line of sight radiating and absorbing at the same wavelength that it is not possible to determine the exact number, since they block each other.

In the limit of small (hv/kT), the exponential term in the Planck function (described in Chapter 3) can be replaced by 1 + (hv/kT) using the approximation e^x = 1 + x. For typical atomic cloud temperatures of 100 K, the exponential term is small (hv/kT = 6.86 × 10^{-4}), so the approximation is valid. The Planck function can be represented by the *Rayleigh-Jeans approximation*:

$$I_\nu = \frac{2kT_B \nu^2}{c^2}$$

The brightness temperature is defined by this equation. Thus, the brightness temperature registered by the radio telescope is directly related to the intensity of the source, which depends on the number of radiating atoms per unit solid angle.

The equation of transfer relates the observed intensity to the intensity coming from a source. There is absorption as well as emission by the gas, so the simple equation of transfer that we wrote in Chapter 4 must be modified to include the emission term if we are observing at a wavelength for which the gas emission can be detected (e.g., 21 cm). The *emissivity*, or energy emitted by the

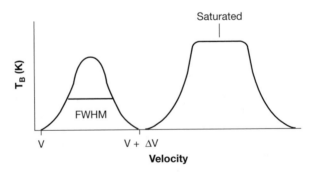

FIGURE 6.2   *The column density is obtained by integrating the emission over the whole spectral line, shown on the left. If the line is saturated, as shown on the right, then the calculated column density is only a lower limit.*

radiation per unit volume per unit solid angle per second, is represented by the *emission coefficient*, j_ν, which is frequency-dependent. The *absorptivity* is represented by the *absorption coefficient*, κ_ν, which is the product of the gas density and effective absorption cross section; the units of κ_ν are inverse length. With this definition, we can rewrite the optical depth as $\tau_\nu = \int \kappa_\nu \, dl$. The *source function*, S_ν, is defined to be the ratio of the emissivity to the absorptivity: $S_\nu = j_\nu/\kappa_\nu$. Then the equation of transfer relates the observed intensity to the emission and absorption properties of the medium and the intensity of the background source by

$$\frac{dI_\nu}{d\tau_\nu} = -I_\nu + S_\nu$$

The derivation of this equation is beyond the scope of this text, but it is detailed in references (e.g., Spitzer 1978, Scheffler and Elsasser 1987).

The gas is said to be in *local thermodynamic equilibrium* when the source function is given by the Planck function for the local gas temperature. If we substitute the brightness temperature for the intensity and integrate the equation of transfer over the optical depth, we get for a uniform medium

$$T_B = T_o(1 - e^{-\tau_\lambda}) + T_b e^{-\tau_\lambda}$$

where T_o is the excitation temperature of the gas, and T_b corresponds to the temperature of a background radio source.

Suppose there is a background radio continuum source along the line of sight to a cloud. Observing in a direction just off the background source at line center frequency makes visible the 21-cm emission from the cloud and its self-absorption along the line of sight; then the equation reduces to

$$T_{B_1} = T_o(1 - e^{-\tau_\lambda})$$

On the background source at a frequency to either side of the hydrogen line, continuum emission will be observed; the equation then reduces to

$$T_{B_2} = T_b$$

These equations, corresponding to observations on and off the background source and at frequencies to either side of line center and at line center, can be used to solve for the optical depth and the gas temperature of the cloud.

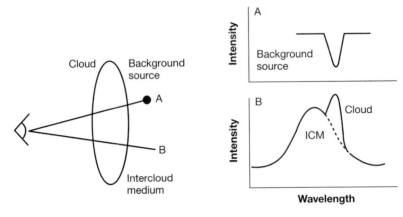

FIGURE 6.3 *(left) Lines of sight through a cloud on and off a background source; (right) the corresponding line profiles.*

The appearance of 21-cm lines is represented in Figure 6.3. The top profile on the right shows absorption of the background continuum 21-cm radiation by a cold cloud. The bottom profile is a line of sight off the background source. The narrow line is from the cold cloud, and the broader shallower line is from a warm intercloud component of the ISM because the line width is proportional to the temperature of the region. The ICM (intercloud medium) does not show up in the absorption profile because the optical depth is proportional to the inverse temperature and so is small for the ICM.

6.2 Radiation from Molecular Gas

Molecules emit radiation as they rotate about their center of mass and oscillate about their average separation of nuclei. These motions are quantized. Molecular hydrogen, H_2, is the most abundant molecule; however, it is difficult to observe because it radiates very weakly at the typically low temperature of the interstellar medium. In general, the strongest radiation from an asymmetric molecule is *electric dipole radiation*, which results from two slightly separated charges with different masses. There is no electric dipole radiation from H_2 because it is a symmetric molecule, so the center of mass and the center of charge coincide. Instead, H_2 radiates weak *quadrupole emission* due to *rotational-vibrational transitions* at high temperatures. It is more practical to examine other molecules that are less abundant but radiate more strongly at low temperatures.

Carbon monoxide is a good tracer of molecular gas. There is only about one ^{12}CO molecule for every 10^4 H_2 molecules. The exact ratio of CO to H_2 varies and is a subject of considerable interest, since observations of CO are critical for determining the total mass content of a region if the gas is optically thin. The great strength of the CO line compensates for its low abundance relative to hydrogen; CO is the most widely observed molecule in our Galaxy and in other galaxies because of its intense 1–0 transition rotation line. The most abundant form of CO consists of the most common isotopes of C and O—that is, ^{12}C and ^{16}O—and is usually just written as ^{12}CO. This molecule has a rotationally excited state that radiates at 2.6 mm from the J = 1 to 0 transition, where J is the *rotational quantum number*. The next excited rotational state leads to a 1.3-mm emission line from the transition J = 2 to 1. The frequency-integrated intensity of the emission is related to the number of molecules radiating per solid angle, so it (indirectly) measures the mass content of a region if the gas is optically thin. There are also other less-abundant isotopes, such as ^{13}CO (a factor of 40 less abundant) and $^{12}C^{18}O$ (a factor of 500 less abundant).

In the solar neighborhood, the column density of CO is empirically related to the temperature (K) and linewidth dv (km s^{-1}):

$$N(H_2 \, cm^{-2}) \approx 2.8 \times 10^{20} \int T_R^*(CO) \, dv$$

The equation is similar to the equation for N(H I). Here T_R^* is the equivalent Rayleigh-Jeans temperature; since the Rayleigh-Jeans approximation is not valid for the 2.6-mm CO line (i.e., $h\nu/kT = 0.55$, which is not <<1), a correction factor relates the excitation temperature of the gas to the Rayleigh-Jeans temperature.

Less abundant molecules are useful for tracing dense regions, because their low abundance is not likely to give a saturated line. The weak line $^{12}C^{18}O$ requires high column densities in order to be observable, whereas CS requires high densities for excitation.

The low temperatures in molecular clouds are a result of *self-shielding* and dust shielding against outside stellar radiation. Typically, a molecular cloud has an overlying atomic layer in which the starlight at the H_2 photodissociation frequencies is absorbed. Cosmic rays are the dominant source of heating in molecular clouds, because they are able to penetrate the clouds; this energy input is balanced by losses from radiating molecules.

6.3 Radiation from Ionized Gas

Interstellar gas may exist in an ionized state at low density and high temperatures. Around O stars, the gas is heated to temperatures between about 7000 K and 14,000 K due to the ultraviolet radiation. This ionized gas is known as an *H II region*; an example in our Galaxy is shown in Figure 6.4. H II regions are also referred to as *Strömgren spheres* after the astronomer who developed the theory to explain the ionization process in the interstellar medium.

In a steady state, the rate of ionization of hydrogen atoms equals the rate of recombination. The number of ionizing photons per unit time, N_{uv} (a function of stellar type), is equal to the integral of the recombination rate, $n^2\alpha$, over the volume of a sphere:

FIGURE 6.4   *H II region M17 in the Milky Way. (Image from the STScI Digital Sky Survey.)*

$$N_{uv} = 4\pi \int r_S^2\, n(r)^2\, \alpha\, dr$$

where α is the *recombination coefficient* for hydrogen and depends on the temperature of the region (details may be found in references such as Osterbrock 1974). The integrated equation may be rewritten to solve for the Strömgren radius, r_s, describing the extent of the ionization:

$$r_S = \left(\frac{3}{4\pi\alpha}\right)^{1/3} N_{uv}^{1/3}\, n^{-2/3}$$

(in this equation, the density is assumed to be constant, although generally it is nonuniform). Different atoms have different ionization energies and recombination coefficients; thus, an H II region may have a Strömgren sphere of ionized helium with a smaller size than the Strömgren sphere of ionized hydrogen.

For purposes of understanding galaxy radiation, one of the key aspects of H II regions is that they produce strong Hα emission (see Chapter 3) because of the recombination and cascade of electrons in hydrogen atoms. Thus, a filter that isolates the Hα line is useful for examining regions of star formation in a galaxy, as shown in Figure 6.5. We will return to this point in Chapter 10 when we examine star formation in more detail.

In addition to the ionized gas around young hot stars, there is a fairly uniform distribution of warm intercloud ionized gas of ~8000 K. There is also a hot component of interstellar matter, with T ~ 10^5 K–10^6 K. This gas is sometimes called the *coronal gas* because its temperature is the same as that of the Sun's corona. Because of its high temperature, the coronal gas has a high kinetic energy. Consequently, it fills a larger volume of galactic space than does cold, low energy gas. The hot component is observed through absorption lines of highly ionized

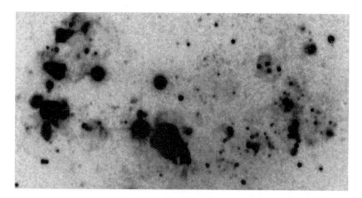

FIGURE 6.5 *Hα image of a portion of the nearby galaxy Holmberg II. Note the shells of ionization that appear as loops. (Imaged with the KPNO 0.9-m telescope*

atoms such as C IV, Si IV, N V, and O VI along the lines of sight to stars. The strengths of these lines increase with the distances to the stars. The highly ionized regions are uniformly distributed but clumpy, with typically three hot gas regions along a 1-kpc line of sight. The hot gas may result from supernova remnants or from thermal evaporation of cold clumpy gas in even hotter material. A model for this distribution is shown in Figure 6.6.

X-ray emission in galaxies is observed in early Hubble types. X-rays radiate from hot gas in the bulge shed by stars; the gas is hot because of the high-velocity dispersion of the bulge. If the velocity dispersion is less than ~200 km s^{-1}, the temperature of the gas falls below 5×10^6 K and will not be detected.

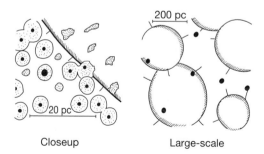

Closeup   Large-scale

FIGURE 6.6 *Distribution of the hot interstellar medium according to the supernovae model (McKee and Ostriker 1977). The hatched areas in both diagrams represent regions of 10^5 K–10^6 K around OB stars and supernovae. (left) A supernova wave is moving across the diffuse clouds, which have 80 K cores (the solid dots); the dotted transition areas are 8000 K envelopes. (right) A larger view: the large circles are supernova remnants, and the small circles are clouds larger than 7 pc.*

6.4 Total Gas Mass

The total gas mass in a galaxy can be determined from the integral of the column density over the area of the galaxy. The column density (number per square centimeter) is converted to a mass column density (grams per square centimeter) by multiplying it by the mass per H atom, 1.67×10^{-24} g (or, for a ratio of H:He = 10:1, the average mass per atom is 2.3×10^{-24} g). The total molecular mass may be similarly estimated on the basis of CO emission. In our Galaxy, the total molecular gas mass is about 4×10^9 $M_\odot$, which is ~2% of the luminous mass, whereas the total H I mass is $5 \times 10^9 M_\odot$.

All spiral galaxies have H I emission, but the amount varies depending on Hubble type and galaxy luminosity. H I has been detected in ~20% of S0s sampled. In these galaxies, the H I gas often is in a ringlike distribution. Sometimes it is concentrated toward the galaxy center, which is very different from the distribution in spiral galaxy disks. Only about 2% of E galaxies have detectable H I, and this gas is probably from the capture of other galaxies or their gas.

An empirical index of H I richness, called the *H I index*, has been developed by de Vaucouleurs:

$$H \text{ I index} = 16.6 - 2.5\log S_H^o - B_T^o$$

where S_H^o is the 21-cm flux density in units of 10^{-28} watts per square meter over all velocities, corrected for self-absorption of 21-cm radiation due to the galaxy's inclination (see RC2); B_T^o is the total apparent B magnitude corrected for inclination (see Chapter 4), and 16.6 is selected so that the H I index is approximately 1 for spirals. A watt is a unit of power, specifically joules per second (a joule = 10^7 erg), so the flux density described here is the same as flux described in Chapter 3, Section 3.1. Thus, the H I index is essentially the ratio (expressed as a magnitude difference) of a galaxy's 21-cm brightness to its optical brightness. The H I index is small for gas-rich galaxies and large for gas-poor galaxies. The index is useful for illustrating the general relation between atomic hydrogen content and Hubble type, as shown in Figure 6.7. The range of H I indices for a given Hubble type is approximately ±0.4.

Some galaxies have very smooth spiral arms that lack bright star-forming regions; van den Bergh developed a classification sequence for these galaxies, which he called *anemic*. His designations Aa through Am are analogous to the Hubble sequence Sa through Sm for normal spirals. The anemics apparently are H I gas-poor for their Hubble type. An example is NGC 4569, shown in Figure 6.8; its H I index is 3.9, compared with the normal value of ~2.7 for this Hubble type.

The amount of CO emission also varies with galaxy type. Elliptical galaxies with strong far-infrared emission have been detected in CO, with inferred gas masses of 2×10^6 to 10^9 $M_\odot$. Field ellipticals have a higher CO detection rate than ellipticals in groups. About 25% of S0 galaxies and 50% of S0a and Sa galaxies have been detected in CO, with corresponding H_2 masses of ~10^7 to 10^9 $M_\odot$.

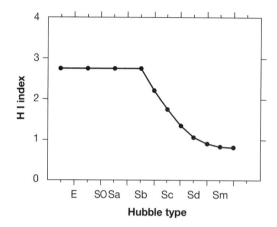

FIGURE 6.7   *Hydrogen abundance relative to blue luminosity as a function of Hubble type, based on the de Vaucouleurs H I index.*

Irregular galaxies are H I-rich but underabundant in CO compared with their CO linewidth. This underabundance could result from less C or less O, from more CO destruction because of relatively less dust than in spirals, or because the CO excitation temperature is less than in spirals.

The gas mass in a galaxy may be compared with the mass of stars. The H I mass to blue luminosity ratio, M_{HI}/L_B, increases from ~0.06 for Sa galaxies to ~0.3 for Sc galaxies. In S0 galaxies, the ratio is $0.03 < M_{HI}/L_B < 1.4$, whereas in elliptical galaxies, which are very gas-poor, the ratio is $M_{HI}/L_B < 0.03\ M_\odot/L_\odot$. The ratio of molecular mass to blue luminosity, $M(H_2)/L_B$, is about the same for Sa through Sc galaxies and decreases by a factor of 3 for Scd through Sdm types.

FIGURE 6.8   *NGC 4192 (left) is an Sb II galaxy; NGC 4569 (right) is an Ab II galaxy. Note the smoothness of the anemic galaxy arms compared with those of the normal spiral. (Images from the STScI Digital Sky Survey.)*

The fraction of the total gas in galaxies (H I + H_2) relative to the total galaxy mass can be expressed as M_{gas}/M_{dyn}. M_{dyn} is the dynamical mass inferred from kinematic measurements; it is determined from the balance between the gravitational force and the centrifugal force from orbital motions (see Chapter 7). The gas fraction ranges from ~4% for Sa galaxies to ~25% for Scd galaxies. Evidently, early-type galaxies have locked up a larger fraction of mass in stars than have late-type galaxies. The mean ratio of the total gas content (H I + H_2) to area inside the Holmberg radius is ~17 $M_\odot/pc^2$, with only a factor of 2 increase from early to late Hubble types; that is, late types have only slightly greater gas surface densities than early types.

The ratio of molecular to atomic mass, H_2/H I, averages 1.0±3.3 for spiral types Sa through Sd. The ratio decreases with later Hubble type: $M_{H_2}/M_{HI} =$ 4.0±1.9 for S0/Sa, and 0.2±−0.1 for Sd/Sm. There is a wide range within each type, as illustrated in Figure 6.9. By comparison, the few ellipticals measured to date have M_{H_2}/M_{HI} ratios of 0.4±0.6.

In general, the phase of the gas varies systematically along the Hubble sequence; later types have lower average molecular gas fractions. Variations in the ratio H_2/H I can result from changes in the efficiency of H_2 cloud formation, the rate of cloud disruption, the rate of molecule formation, or the metallicity.

There is an increase in H I and H_2 masses with increasing dust content in galaxies; typically the gas-to-dust mass ratio is ~100. IRAS observations of radiation at 100μ are used to infer dust masses. The dust mass per blue luminosity, M_{dust}/L_B, is a factor of 10 larger in Sa compared with S0a galaxies. Recent studies indicate that there may also be cold dust that does not radiate at wavelengths detectable by IRAS; as much as 90% of dust may be hidden in this way. Ellipticals have gas-to-dust mass ratios as high as 700, probably due to cold dust.

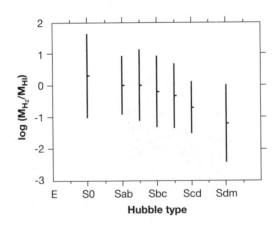

FIGURE 6.9   *Ratio of molecular to atomic gas mass as a function of Hubble type. (Based on Young and Scoville 1991.)*

6.5 Distribution of Clouds

We will now examine the distributions of the different phases of the interstellar medium in galaxies. Atomic clouds typically have densities of a few tens of atoms per cubic centimeter and temperatures of ~100 K. There is a large range of cloud sizes; the smallest observed atomic clouds are *diffuse clouds,* which are a few parsecs in size with ten to a few hundred solar masses. Diffuse clouds are spread rather uniformly throughout the midplane of the disk, with a vertical distribution spanning about 100 pc. The largest atomic clouds have masses ~10^6–10^7 M_o and 0.5 kpc sizes. They are sometimes regularly spaced along spiral arms, with separations that scale with the size of a galaxy, typically ~$0.2 R_{25} = 1$–4 kpc.

The smallest isolated molecular clouds are *Bok globules*, which are smaller and denser than diffuse atomic clouds of the same mass (see Figure 6.10). Slightly larger clouds are referred to as *dark clouds*; the largest aggregates are *giant molecular clouds* (GMCs), with masses up to 10^6 M_o and sizes up to ~100 pc. Average densities are ~100 per cubic centimeter in molecular clouds, increasing to ~10^5 per cubic centimeter or more in the dense cores. Figure 6.11 shows a CO intensity contour map in the region W3/W4/W5 in the Milky Way.

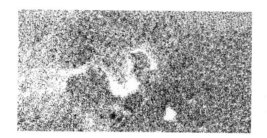

FIGURE 6.10 *A filamentary Bok globule, Lynds 66, commonly known as the Snake. (Image from Schneider and Elmegreen 1979.)*

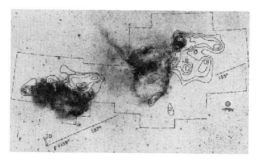

FIGURE 6.11 *Molecular clouds in W3/W4/W5 in the Milky Way, as traced by CO temperature contours. (From Lada et al. 1978.)*

CO observations in our Galaxy indicate that massive clouds may contain most of the molecular gas, even though they are few in number compared with smaller clouds. About 40% of the molecular mass in the galactic disk is in complexes with masses between 10^4 $M_\odot$ and 5×10^6 $M_\odot$ based on observations in the first galactic quadrant (i.e., between galactic longitudes 0° and 90°; see Chapter 8). The distribution can be described by a *mass function*, or number of clouds N in a range M + dM as a function of mass. The distribution follows a power law, given by

$$N(M)dM \sim M^{-1.5}dM$$

for clouds with mass M > 3×10^5 $M_\odot$. The observed mass functions are shown in Figure 6.12.

Clouds appear to be hierarchically clumped, or fractal. Figure 6.13 shows a model fractal cloud.

Several dozen giant cloud complexes with masses >10^5 $M_\odot$ have been detected in quadrant I in the Milky Way (i.e., between galactic longitudes 0° and 90°; see Chapter 8). These complexes tend to be in the main optical spiral arms; their distribution is shown in Figure 6.14.

Irregular (LMC-type) galaxies also contain 10^6–10^7 $M_\odot$ H I clouds. The number is limited by galaxy size: Type Im galaxies have less than 10 giant H I clouds, whereas spiral galaxies have several hundred.

Spiral and irrregular galaxies often contain giant H I holes several hundred parsecs in size, as shown in Figure 6.15. They evidently are caused by the com-

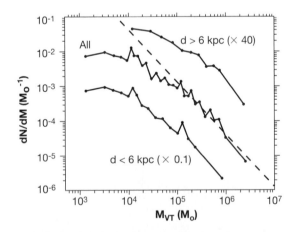

FIGURE 6.12 *The mass distribution function for molecular clouds, based on CO observations in our Galaxy in the longitude range $8° < l < 90°$. The dashed line is a least-squares fit to the data for clouds more massive than 3×10^5 $M_\odot$. The lower and upper functions are for clouds at distances <6 kpc and >6 kpc, respectively. (From Solomon and Rivolo 1989.)*

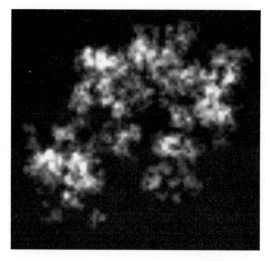

FIGURE 6.13   *Fractal model for the interstellar medium. (Based on computer simulations by B. Elmegreen, 1997, IBM Watson Research Center.)*

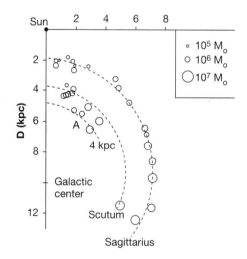

FIGURE 6.14   *Schematic diagram of the CO distribution from giant molecular clouds in quadrant I of our Galaxy. (From Dame et al. 1986.)*

bination of supernovae, photoionization, and stellar winds that are energetic enough to ionize and blow out the local material perpendicular to the disk. In some spirals, holes may also be caused by high velocity clouds impacting the disk. The holes are filled or outlined by Hα emission due to star formation.

Table 6.1 summarizes the properties of different types of clouds, and Figure 6.16 shows a diagram of the different phases. Note that there is an approximate thermal pressure balance between the lowest-density parts of the different

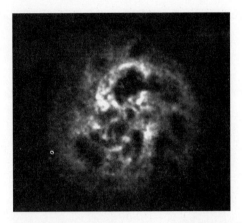

FIGURE 6.15   *H I holes in the irregular galaxy Holmberg II. (Based on 21-cm Very Large Array observations by Puche et al. 1992; image from D. Westpfahl.)*

TABLE 6.1   *Properties of interstellar clouds*

Type	Size (pc)	T (K)	n (cm^{-3})	M (M$_o$)
Diffuse (H I)	10s	100	~10	100s
Dark (H$_2$)	10s	10	~100–1000	100s
Giant molecular cloud (H$_2$)	100s	10	~100 (10^5 in cores)	10^4–10^5
Giant atomic (H I)	1000s	100	~10	10^6–10^7

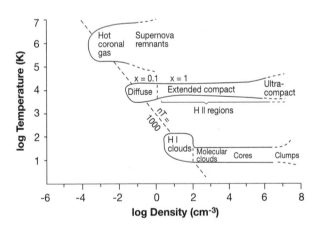

FIGURE 6.16   *Phases of the interstellar medium. (From Yorke 1987.)*

components of the interstellar medium: the pressure, P, is given by $P = nkT$ for density n and temperature T.

Because gas and dust are mixed together, there is a linear relationship between the column density of gas and the amount of extinction. From a comparison of 21-cm observations and reddening along an average line of sight, the relation is

$$N(H\ atoms/cm^2) = 1.9 \times 10^{21} A_v\ (mag)$$

Thus, for every magnitude of visual absorption along a line of sight, there are about 2×10^{21} hydrogen atoms per square centimeter. This relationship comes from observations of color excess E(B-V) converted to magnitudes of absorption through the standard reddening law

$$R \equiv \frac{A_v}{E(B - V)} = 3.1$$

At high opacities, atoms may form molecules. Observations with the Orbiting Astronomical Observatory-2 (OAO-2, also known as "Copernicus"), taking into account both atomic and molecular gas along an average line of sight, give the relation

$$N(H\ I) + 2N(H_2)\,(atoms/cm^2) = 5.8 \times 10^{21}\ E(B - V)\ (mag)$$

for the average diffuse interstellar medium in our galaxy, which may be nonlinear at very high extinctions. Figure 6.17 shows the relationship between the total hydrogen (H I + H_2) and color excess.

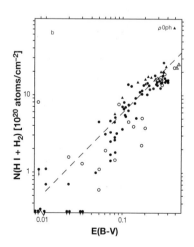

FIGURE 6.17 *Total hydrogen as a function of color excess in the Milky Way. (From Bohlin et al. 1978.)*

A detailed map of the reddening in the Milky Way for different longitudes and latitudes has been made by D. Burstein and C. Heiles using galaxy counts and 21-cm observations compared with color excess.

6.6 Radial Gas Density Profiles

In general, the atomic hydrogen distribution is flat over most of the visible disk of a galaxy, decreasing with increasing radius in the outer parts. The radial distribution of H I in the Milky Way midplane has a hole out to 4 kpc, is flat between 5 and 13 kpc, and decreases out to about 18 kpc, as shown schematically in Figure 6.18.

The gas density n(H I) declines steeply after the Holmberg radius R_H (where the surface brightness is 26.5 mag arcsec^{-2}) but can be seen beyond R_H for most galaxies; sometimes it can be detected to $2 R_H$ or so. In some galaxies, H I clumps have been detected as far out as 100 kpc from the nucleus; these appear to be high-velocity clouds resulting from tidal streaming. Some of these in other galaxies may account for absorption lines seen on the lines of sight to quasars (see Chapter 11). No other gas components besides H I have a flat distribution with radius.

In general, the molecular gas content decreases exponentially with radius. Figure 6.18 shows the radial distribution of CO in the Milky Way. The radial distributions along the major axes of several Sc galaxies are shown in Figure 6.19. The Milky Way profile has a central CO peak, a decrease of gas between 1 and 4 kpc, and a molecular annulus (ring) between 4 and 8 kpc; beyond that, the profile decreases exponentially. The Sc galaxy IC 342 has a similar (but shallower) dip inside 4 kpc and a peak at about 6 kpc. The other galaxies in this figure have a central peak with no dip. The radial profiles of CO in earlier-type spirals are more difficult to observe, because the gas content is lower in these galaxies. Less than

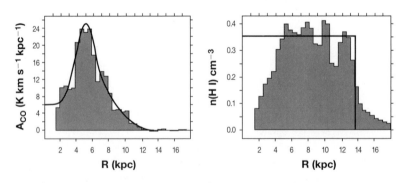

FIGURE 6.18   *Intensity of CO (left) and density of H I (right) as a function of radius in the Milky Way. The lines are smoothed approximations to the histograms. (From Burton and Gordon 1978.)*

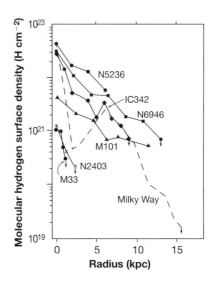

FIGURE 6.19 *Radial distribution of H_2 in Sc galaxies and the Milky Way based on CO observations. (From Young and Scoville 1991.)*

half of the measured Sb and Sbc galaxies have CO depressions in their centers. Beyond the central region, Sb, Sbc, and Sc galaxies all have declining CO intensity with radius. The radial CO distributions have been measured in only a small sample of Sas because of the difficulty of detecting their weak CO radiation; their CO radiation also declines with galactocentric distances. Most of these Sas do not have central CO holes.

Holes in one gas phase sometimes are accompanied by an increase in the other phase, and sometimes there is an absence of gas. For example, M33 has no H I or CO hole, whereas M81 has both a central H I hole and a central CO hole. In N6946, there is a central CO peak and an H I hole; in this case, the central H I hole evidently exists because the gas in the center is mostly molecular.

In the Milky Way clouds, the fraction of gas in the form of molecules depends on the distance from the Galactic Center. On the basis of the CO mass of giant molecular clouds compared with their associated giant H I clouds, the molecular fraction in individual clouds decreases with radial distance from ~70% at 4 kpc to 5% at 9 kpc, as shown in Figure 6.20.

The observed radial decrease in the ratio H_2/H I in galaxies implies that molecular cloud formation is most efficient near the center. Inefficient molecular cloud formation in the outer parts of a galaxy could result from decreasing gas volume densities as the H I scale height increases. Disk pressure also affects molecule formation; higher pressures lead to higher densities, which increase collision rates between atoms and dust grains. This increase, in turn, facilitates H_2 formation. Higher densities also shield the molecules from radiation, which would destroy their bonds.

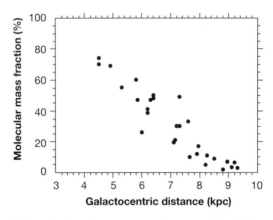

FIGURE 6.20   *Molecular fraction versus galactocentric distance. (Based on Elmegreen and Elmegreen 1987.)*

6.7 Metallicities in the Milky Way and Other Galaxies

It is conventional in astronomy to refer to all atoms besides hydrogen and helium as metals, and their abundances relative to hydrogen as their *metallicities*. Metallicities vary from one galaxy to another, but spirals and ellipticals with the same luminosity have similar metallicities. This result is perhaps surprising, since ellipticals and spirals have very different star formation histories. In the first billion years after most galaxies were formed, star formation was more rapid in ellipticals than in spirals or irregulars. As a result, elliptical galaxies have less gas today than spirals or irregulars; the low mass stars are still slowly evolving, but essentially no new star formation is occurring. Spirals and irregulars are forming high mass stars rapidly today because they have large masses of gas left over from galaxy formation. This scenario suggests that the earlier star formation in ellipticals accounts for their observed metallicities, whereas the more recent star formation in spirals accounts for their metallicities. Some astronomers think that the different star formation rates are related to the absence or presence of neighboring galaxies, which perturb the gas in a galaxy. Star formation properties of galaxies will be considered in more detail in Chapter 10, and more about environmental effects will be considered in Chapter 11.

In galaxies that can be observed with sufficient sensitivity, metallicities are determined by measuring emission lines from H II regions or interstellar clouds, or by measuring the strengths of lines in stellar atmospheres. In the Milky Way, we can also examine absorption lines in the interstellar medium caused by clouds along the lines of sight to stars. The main difficulty in determining the abundance of an element in a galaxy is in accounting for all possible states (ionized, atomic, molecular, or condensed onto grains). For example, in the case of stellar atmospheres, singly ionized lines of carbon are commonly observed. Estimates for the expected amount of higher ionizations must be made in order to derive meaningful abundances. In clouds, carbon can exist as a neutral atom, an ionized atom, or any of a wide variety of molecules. Typically about 12% of carbon in molecules is in the form of CO. Dust grains also contain carbon.

It is customary to define metallicity by the ratio of the density of an element relative to hydrogen compared with the solar ratio. For example, the metallicity based on iron in a star is given by

$$\left[\frac{Fe}{H}\right] \equiv \log\left[\frac{n(Fe)}{n(H)}\right]_* - \log\left[\frac{n(Fe)}{n(H)}\right]_o$$

for density, n, in a stellar atmosphere, designated by subscript $*$, relative to the solar value, designated by subscript o. Note from this definition that if [Fe/H] = 0, the abundance is the same as the solar value.

The Hyades cluster in our Galaxy (see Figure 6.21) is a useful observational reference for [Fe/H]. Its value is 0.025, giving a metallicity compared with the Sun of $10^{0.025} = 1.059$, so it is nearly the same as the solar value.

Measurements have been made of the variation of [Fe/H] in stars as a function of position in the Galaxy for different spectral types. Overall, the ratio near the bulge is [Fe/H] = 1, decreasing to -2.3 for the most metal-poor globular clusters. In the disk, the average [Fe/H] is -0.5 for the thin disk, -0.8 to -0.6 for

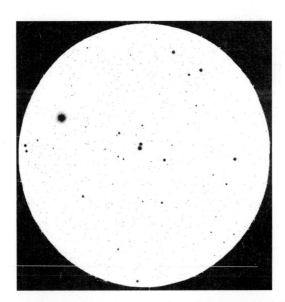

FIGURE 6.21   *The Hyades open cluster in the Milky Way is very diffuse and spans about 5° on a side. The brightest star, on the left, is Aldebaran and is not part of the Hyades. The other bright stars are part of the cluster, and the V-shaped head of the constellation Taurus is visible. (Image from Vehrenberg and Guntzel-Lingner 1983.)*

the thick disk, and < -1.6 for the halo component. There is a radial decrease in metallicity by a factor of 2 to 3 from the Galactic Center outward to 10 kpc; the metallicity of young stars decreases with distance R from the galactic center more rapidly than that of old stars. In general, the gradient d[Fe/H]/dR $\approx$ -0.08/kpc for young F dwarfs and G or K giants, and -0.02 to -0.03 per kiloparsec for K giants and intermediate-age FGK stars. The gradient is steeper in the outer disk: for K giants, d[Fe/H]/dR ≈ -0.05 kpc at R = 10 kpc, and -0.1/kpc for R > 10 kpc. Beyond 10 kpc, there seems to be no gradient in the spherical component but a constant abundance ratio of [Fe/H] $\approx -2.2\pm0.3$, which is a factor of 160 less than the solar value.

Radial metallicity gradients are also observed in other galaxies, where it is easier to sample over the whole disk than in our galaxy. The gradients may be related to radial star formation rates, as discussed further in Chapter 10. The gradients are less steep in barred spirals than in nonbarred spirals, presumably because bars cause radial mass motions, which then mix the metallicities.

There is also an abundance gradient for stars perpendicular to the plane of a galaxy. For example, there appears to be a steep gradient, d[Fe/H]/dz ≈ -0.5 to -0.6/kpc (a factor of 3–4/kpc), among Population I stars in our galaxy. In stars, metallicity can be determined from ultraviolet photometry. Recall from Chapter 3 that a star of a given spectral type has specified (U-B) and (B-V) colors; sometimes the (U-B) color is higher than it should be for the star's type. The photometric index of *uv excess*, δ(U-B), measures the extra (U-B) light from a star relative to the expected (U-B) based on its (B-V). This uv excess is well correlated with the metallicity [Fe/H], as seen in Figure 6.22.

This figure shows that regions with a higher uv excess (positive values on the ordinate) have a smaller metallicity (which is at the right end of the plot); for

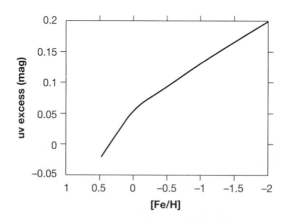

FIGURE 6.22 *Ultraviolet excess versus metallicity, which decreases toward the right. The uv excess is higher when the metallicity is lower, because then there is less line blanketing.*

example, a uv excess of 0.2 mag corresponds to a metallicity that is a factor of 100 less than the solar value. This relation occurs because strong absorption lines make the uv part of the spectrum less bright, a phenomenon referred to as *line blanketing*. If there is a lower metallicity, the star will appear brighter than normal in the uv because of weaker absorption lines. The relation is [Fe/H]-[Fe/H]$_{Hyades}$ ≈ 5 δ(U–B).

In our Galaxy as well as other galaxies, bright emission lines from ionized gas in H II regions are useful for probing conditions in the gas. Some ratios of line strengths are good temperature indicators; others are good density indicators. For example, the forbidden doubly ionized oxygen lines [O III] are temperature-sensitive because they involve transitions between either of two upper levels and one split lower level; the temperature determines the relative populations of the upper levels, as determined from the Boltzmann equation described in Chapter 3. The upper transition produces radiation at λ = 4363 Å, and the lower transition produces radiation at λ = 4959 Å and 5007 Å into the split ground state. The ratio is

$$\frac{I_{4959} + I_{5007}}{I_{4363}} = \frac{8.32 \exp\left(\dfrac{3.29 \times 10^4}{T}\right)}{1 + \left(4.5 \times 10^{-4}\dfrac{n_e}{T^{1/2}}\right)}$$

for gas temperature T and electron density n_e, as shown in Figure 6.23. The coefficients depend on the collision strengths and transition probabilities, as

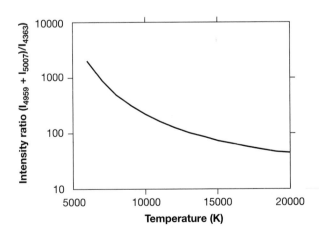

FIGURE 6.23 *Intensity ratios for [O III] lines as a function of temperature, assuming the low density limit of n_e↦0. The limit on the right approaches the ratio of the statistical weights of the two [O III] lines.*

detailed by Osterbrock (1974). The electron configuration is the same in [N II] as in [O III], with radiation at 5755 Å from the upper level and at $\lambda = 6548$ Å and 6583 Å into the lower levels.

The sum of the intensities of the two strong lines ($I_{4959} + I_{5007}$) is often compared with a reference line, Hβ, because it is strong and has a convenient nearby wavelength of 4861 Å. The log of the ratio, written as log {[O III]/Hβ}, is known as the *excitation* of an H II region. The excitation increases with increasing radius in Sc galaxies, as shown in Figure 6.24.

The ratio of the forbidden singly ionized oxygen lines [O II] at $\lambda = 3726$ Å and 3729 Å is a density indicator; the temperature required to populate each of two split upper levels is approximately the same, but the collisional excitation rates to these two levels are different. At low density, the ratio of line strengths equals the ratio of collisional excitation rates, since each collisional excitation is followed by a spontaneous radiative decay. At high density, the levels are in thermal equilibrium with the gas and the line radiation depends only on the relative rates of spontaneous decay, independent of the collision rate. Between these two limits, the line ratio varies systematically with density. This means that the relative line strengths can be used to infer the density.

The ratio log [O II/N II] also increases with an increasing ratio log {[O III]/Hβ}. Because more energy is required to excite O II than N II, this correlation is consistent with higher excitation at larger galactocentric distances. One interpretation of these results is that the excitation is low in H II regions at small galactocentric distances because the overall metallicity is higher there: The presence of more metals leads to more cooling due to the escaping radiation from the deexcitations of the ions. Consequently, the observed [O III] lines are stronger with decreasing oxygen. The inferred relative oxygen abundance log [O/H] decreases approximately exponentially with radius by a factor of 10 from the center to about 0.5 R_{25}, and the ratio log [N/O] decreases slightly with increasing radius.

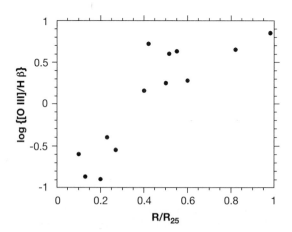

FIGURE 6.24   *[O III]/Hβ ratio versus radius in Sc galaxies. (Adapted from Searle 1971.)*

One explanation of these results is that OB star temperatures vary from the center of a galaxy to the edge, and the variation affects the H II region excitations. Another explanation is that the Hβ line strength in H II regions varies with galactocentric radius because of a changing dust/gas ratio. Because a radial variation in the line strengths is seen for many elements, the interpretation that these are abundance gradients is favored.

There are no gradients in the ratios of Ne/O or S/O, presumably because Ne, S, and O are synthesized in coincident sites. Use of the ratio [S II]/[S III] together with [O II]/[O III] allows a determination of stellar temperature and ionization parameters that is essentially independent of metallicity. Overall, the mean metallicities of disk galaxies increase with increasing maximum rotational velocity (see Chapter 7), increasing luminosity, and decreasing Hubble type, although there is a large intrinsic dispersion in the mean abundances for each type.

Metallicities in ellipticals must be studied by measuring lines in their globular cluster stars, because the abundance is too low to measure easily in the gas. Ellipticals have total metallicities that increase with increasing galaxy luminosity; [Fe/H] varies between 0.2 and -0.2 for visual magnitudes -22 to -17.

The magnesium index Mg_2 is also a good measure of metallicities, and it scales with the ratio [Fe/H]. The index is defined as the difference in magnitude between the flux in a passband centered on the magnesium feature at ~5200 Å and the continuum interpolated from two passbands on either side of the magnesium line. The Mg_2 index increases with increasing elliptical galaxy absolute magnitude.

The bulges of spiral galaxies have metallicities that are very similar to those of ellipticals, and their Mg_2 index is well correlated with the bulge absolute magnitude, as shown in Figure 6.25. There is an even tighter correlation between the Mg_2 index and the central velocity dispersion (see Chapters 7 and 8).

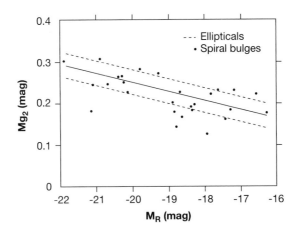

FIGURE 6.25 *The Mg_2 index scales with the absolute R magnitude of elliptical galaxies (solid line, with dispersion indicated by dashed lines) and spiral galaxy bulges (dots). (From Jablonka et al. 1996.)*

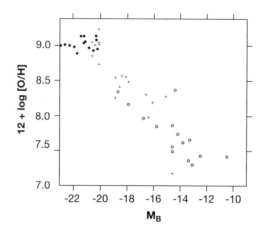

FIGURE 6.26   *Oxygen abundance varies with galaxy absolute magnitude.*
(Based on Roberts and Haynes 1994.)

A few dwarf Es have been found with higher metallicities than large Es, which suggests that these dEs formed from metal-rich galaxies (for example, from mergers; see Chapter 12).

Ellipticals that are in denser galaxy neighborhoods have more metals than do ellipticals that are relatively isolated. This result is consistent with different star formation rates or with different formation mechanisms for galaxies in different environments, as we will discuss more in Chapter 11. Studies of M32, the dwarf E companion to M31, as well as small ellipticals in the Virgo cluster indicate that the metallicity decreases with distance from the cluster center.

Abundances are strongly correlated with galaxy mass, using either luminosity (starlight) or velocity (see Chapter 7) as mass indicators. Figure 6.26 shows a linear relation between oxygen abundance and absolute magnitude for spiral galaxies, over an interval covering several magnitudes and a factor of 300 in abundance. Essentially the same correlation has been found for elliptical, lenticular, and irregular galaxies as well. These findings suggest that mass is the key factor that determines a galaxy's metallicity. This result is consistent with the color-magnitude effect mentioned in Chapter 4; brighter (presumably more-massive) galaxies of all types are redder because of higher metallicities, because metals decrease the blue light emission.

Exercises

1. Observations of G and K dwarfs in the vicinity of the Sun yield a radial iron-to-hydrogen gradient of $d[Fe/H]/dR = -0.05$/kpc. Determine the iron-to-hydrogen abundance ratio at a distance of 3 kpc beyond the Sun; express your answer as a percentage.
2. In our Galaxy, the helium-to-hydrogen abundance gradient $d[He/H]/dR$ as measured through planetary nebulae observations is -0.02/kpc; for oxygen-to-hydrogen it is

d[O/H]/dR = −0.06/kpc. By what factors do the [He/H] abundance and the [O/H] abundance decrease at 10 kpc compared with 5 kpc?

3. It is customary to write abundances in the form {12 + log n(x)/n(H)} for the density of element x relative to hydrogen; thus, if iron has n(Fe)/n(H) = 3.4 × 10⁻⁵, the abundance is expressed as 12 + log (3.4 × 10⁻⁵) = 7.53. M. McCall et al. (1985) studied several nearby galaxies and found a correlation between the ([O II]+[O III])/Hβ ratio and the abundance of oxygen, such that a plot of {12 + log [n(O)/n(H)]} (y-axis) versus log {([O II]+[O III])/Hβ} (x-axis) has points (x,y) at (0.35,9.0) and (0.65,8.5). Using this information, determine the abundance of oxygen relative to hydrogen (expressed as a percentage) if the observed oxygen fine structure lines relative to Hβ yield a value of log {([OII]+[OIII])/Hβ} = 0.5.

4. The metallicity-I band surface brightness models by S. Ryder (1995, *Astrophysical Journal* 444:610) provide a good fit to observational data with the approximate relation {12 + log [n(O)/n(H)]} = −0.25 mag arcsec⁻² + 14.25. Determine [O/H] (expressed as a percentage) for a galaxy with an I band surface brightness of 22 mag arcsec⁻².

5. Show that the Rayleigh-Jeans approximation is not valid for a molecular cloud.

6. A saturated ¹²CO line has a peak observed temperature of 20 K and a linewidth of 3 km s⁻¹. Determine the minimum H_2 column density and visual extinction.

7. The [O III] lines at λ = 4959 Å and 5007 Å are observed to be 330 times brighter than the λ = 4363 Å line. If the electron density is 95 cm⁻³, find the temperature of the nebula.

8. Determine the (a) optical depth, (b) column density, and (c) temperature of an H I cloud observed by H I absorption. The brightness temperature on source and on line center is 160 K; the brightness temperature on the cloud but to the north of the background continuum source is 25 K, to the east is 28 K, to the south is 30 K, and to the west is 26 K. The brightness temperature of the background radio continuum source off 21 cm is 180 K. Assume that Δv is given by √(2πkT/m_H) in cgs units.

Unsolved Problems

1. What is the chemical evolution of the Milky Way and other galaxies?
2. What is the intrinsic structure of interstellar clouds, and how much do we miss by low resolution limits?

Useful Websites

Simulations and observations of molecular cloud structures are available on
    http://www-astro.phast.umass.edu/~brunt/fake.html

Further Reading

Bohlin, R., B. Savage, and J. Drake. 1978. A survey of interstellar H I from Lα absorption measurements: II. *Astrophysical Journal* 224:132.

Burstein, D., and C. Heiles. 1982. Reddenings derived from H I and galaxy counts: Accuracy and maps. *Astronomical Journal* 87:1165.

Burton, W., B. Elmegreen, and R. Genzel. 1992. In *The galactic interstellar medium: Saas-Fee advanced course 21*, ed. D. Pfenniger and P. Bartholdi. New York: Springer-Verlag.

Burton, W., and M. Gordon. 1978. Carbon monoxide in the galaxy. III. The overall nature of its distribution in the equatorial plane. *Astronomy and Astrophysics* 63:7.

Buzzoni, A., G. Gariboldi, and L. Mantegazza. 1992. The magnesium Mg_2 index as an indicator of metallicity in elliptical galaxies. *Astronomical Journal* 103:1814.

Dame, T., B. Elmegreen, R. Cohen, and P. Thaddeus. 1986. The largest molecular cloud complexes in the first galactic quadrant. *Astrophysical Journal* 305:892.

Dickman, R., R. Snell, and J. Young. 1988 *Molecular clouds in the Milky Way and external galaxies.* New York: Springer-Verlag.

Efremov, Y. 1995. Star complexes and associations: Fundamental and elementary cells of star formation. *Astronomical Journal* 110:2757.

Elmegreen, B., and D. Elmegreen. 1987. H I superclouds in the inner galaxy. *Astrophysical Journal* 320:182.

Elmegreen, B., M. Morris, and D. Elmegreen. 1980. On the abundance of carbon monoxide in galaxies: A comparison of spiral and Magellanic irregular galaxies. *Astrophysical Journal* 240:455.

Elmegreen, D., and B. Elmegreen. 1979. CO and near-IR observations of a filamentary cloud, L43. *Astronomical Journal* 84:615.

Garnett, D. 1993. H II regions as probes of galaxy evolution and the properties of massive stars. *Publications of the Astronomical Society of the Pacific* 105:996.

Huang, J.-S., A. Songaila, L. Cox, and E. Jenkins. 1995. Detection of hot gas in the interstellar medium. *Astrophysical Journal* 450:163.

Jablonka, P., P. Martin, and N. Arimoto. 1996. The luminosity-metallicity relation for bulges of spiral galaxies. *Astronomical Journal* 112:1415.

Lada, C., B. Elmegreen, H.-I. Cong, and P. Thaddeus. 1978. Molecular clouds in the vicinity of W3, W4, and W5. *Astrophysical Journal Letters* 226:L39.

Lynds, B.D. 1962. Catalog of dark nebulae. *Astrophysical Journal (Suppl.)* 7:1.

McCall, M., P. Rybski, and G. Shields. 1985. The chemistry of galaxies. I. The nature of giant extragalactic H II regions. *Astrophysical Journal (Suppl.)* 57:1.

McKee, C., and J. Ostriker. 1977. A theory of the interstellar medium—Three components regulated by supernova explosions in an inhomogeneous substrate. *Astrophysical Journal* 218:148.

Osterbrock, D. 1974. *Astrophysics of gaseous nebulae.* San Francisco: W.H. Freeman.

Puche, D., D. Westpfahl, E. Brinks, and J.-R. Roy. 1992. Holmberg II: A laboratory for studying the violent interstellar medium. *Astronomical Journal* 103:1841.

Rana, N. 1991. Chemical evolution of the galaxy. *Annual Review of Astronomy and Astrophysics* 29:129.

Roberts, M., and M. Haynes. 1994. Physical parameters along the Hubble sequence. *Annual Review of Astronomy and Astrophysics* 32:115.

Scheffler, H., and H. Elsasser. 1987. *Physics of the galaxy and interstellar matter.* New York: Springer-Verlag.

Schneider, S., and B. Elmegreen. 1979. A catalog of dark globular filaments. *Astrophysical Journal (Suppl.)* 41:87.

Searle, L. 1971. Evidence for composition gradients across the disks of spiral galaxies. *Astrophysical Journal* 168:327.

Solomon, P., and M. Edmunds. 1980. *Giant molecular clouds in the galaxy: Proceedings of the Third Gregynog Astrophysics Workshop*, Cardiff: University of Wales.

Solomon, P., and A. Rivolo. 1989. A face-on view of the first galactic quadrant in molecular clouds. *Astrophysical Journal* 339:919.

Spitzer, L. 1978. *Physical processes in the interstellar medium.* New York: Wiley.

van den Bergh, S. 1991. What are anemic galaxies? *Publications of the Astronomical Society of the Pacific* 103:390.

Vehrenberg, H., and U. Guntzel-Lingner. 1983. *Atlas of deep-sky splendors.* Cambridge: Cambridge University Press.

Wiklind, T., F. Combes, and C. Henkel. 1995. The molecular cloud content of early-type galaxies. V. CO in elliptical galaxies. *Astronomy and Astrophysics* 297:643.

Wiklind, T., and C. Henkel. 1989. The molecular cloud content of early-type galaxies. I. Detections and global properties of ellipticals and lenticulars. *Astronomy and Astrophysics* 225:1.

Yorke, H. 1987. *Radiation in moving gaseous media*, ed. Y. Chmielowski and T. Lanz, 193. Geneva: Geneva Observatory.

Young, J. 1988. The molecular content of galaxies as a function of luminosity. In *Molecular clouds in the Milky Way and external galaxies*, ed. R. Dickman, R. Snell, and J. Young. 326. New York: Springer-Verlag.

Young, J., and N. Scoville. 1991. Molecular gas in galaxies. *Annual Review of Astronomy and Astrophysics* 29:597.

Young, J. et al. 1995. The FCRAO extragalactic CO survey. I. The data. *Astrophysical Journal (Suppl.)* 98:219.

Zaritsky, D. 1992. The radial distribution of oxygen in disk galaxies. *Astrophysical Journal Letters* 390:73.

Zaritsky, D., R. Kennicutt, and J. Huchra. 1994. H II regions and abundance properties of spiral galaxies. *Astrophysical Journal* 420:87.

CHAPTER 7

Rotation in Galaxies

Chapter Objectives: to understand the kinematic behavior of galaxies and how it is used to measure mass

---

Toolbox:

rotation curves	Tully-Fisher relation
kinematic axes	fundamental plane
deprojection	mass-to-light ratio
power law fits	dark matter

7.1 Doppler-Shifted Motions in Spiral Disks

Doppler-shifted spectral lines from optical and radio emission can be used to study motions in the disks of spiral galaxies. Recall that the Doppler formula for nonrelativistic motions ($v_r \ll c$ for radial velocity v_r) is:

$$\frac{\Delta\lambda}{\lambda} = \frac{v_r}{c}$$

The radial velocity that is measured at a given point in a galaxy depends on the internal motions in that disk plus the inclination of the galaxy to our line of sight. The average radial velocity of a galaxy is called the *systemic velocity*. It is a combination of the Doppler-shifted velocity from the general expansion of the Universe (see Chapter 4), random motions of that galaxy from other perturbing galaxies, and solar motion. Note that the shift applies only to the component of motion along our line of sight, so that a source moving at some angle relative to us will show a Doppler shift for just the radial component and not the tangential component of motion. Thus, there is very little net Doppler shift on the minor axis of a galaxy except for the systemic velocity, since there the disk motions are nearly perpendicular to our line of sight for circular motions. The velocity along the major axis gives the projected rotation curve (velocity as a function of galactocentric distance) of the galaxy and tangential peculiar motions.

In general, we do not see rotational motions for a face-on galaxy, so a galaxy must be fairly highly inclined for good kinematic information. However, morphology is more difficult to discern in an inclined galaxy. In practice, a minimum of about 45° is useful for kinematic studies, and a maximum of about 60° is useful for morphological studies.

Mapping the *velocity field* in a galaxy requires observations at many points in the disk. The resulting velocities must be corrected for inclination. The circular velocity in the plane of the galaxy, $v_{circ}(r,\theta)$, is related to the observed velocity, $v_{obs}(s,\alpha)$, by the equation:

$$v_{circ}(r,\theta) - v(0) = \frac{v_{obs}(s,\alpha)\sqrt{\cos^2 i + \tan^2\alpha}}{\sin i \cos i}$$

for systemic velocity $v(0)$, inclination i, and angle α measured counterclockwise between the major axis and the observed point, as shown in Figure 7.1. The angle θ in the disk of the galaxy is related to the angle α by $\tan\theta = \tan\alpha/\cos i$.

The true distance r between the center and any point in a galaxy is determined from the observed distance s between the center and the point by correcting for projection:

$$r = s\sqrt{1 + \sin^2\alpha \tan^2 i}$$

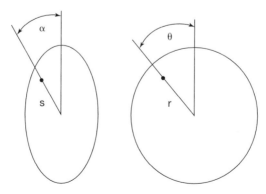

FIGURE 7.1   *(left) Sky view of an inclined galaxy, as it appears to us. (right) Deprojected view (in the plane of the galaxy). The point indicated by a dot appears to be a distance s from the center, but is really a distance r from the center. The angles are measured counterclockwise from the major axis.*

These relations can be inverted to show the expected radial velocity that would be observed at any point (s,α) for a perfectly symmetric inclined galaxy. For example, the inverted relation can be used to determine the projected circular velocity so that it can be subtracted from an observed velocity to get residual velocities such as expected from spiral density wave theory (see Chapter 9). The inverted relation is:

$$v_{obs}(r,\theta) = [v(r) - v(0)]\cos\theta \sin i$$

The radial coordinate in the inclined galaxy is

$$s = r\sqrt{\cos^2\theta + \sin^2\theta \cos^2 i}$$

and the angular coordinate is given by $\tan \alpha = \tan \theta \cos i$.

In order to understand how the distribution of radial velocities appears to us, consider an edge-on view of uniform circular motion in a disk. The observer, at the bottom of Figure 7.2, sees redshifted emission on the left and blueshifted on the right for a galaxy rotating clockwise. Objects that are different distances from the center move at different angular rates with respect to the center.

Now imagine that the galaxy is inclined to our line of sight. The three points connected by an arc in Figure 7.3 all have the same observed radial velocity. The arc has coordinates (s,α) that satisfy the equation v_{obs} = constant for v_{obs} given above.

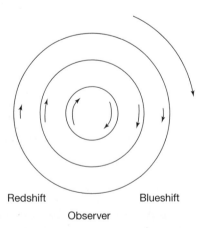

FIGURE 7.2   *Galaxy rotation is observed as redshifts and blueshifts.*

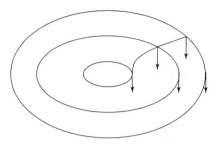

FIGURE 7.3   *The arc connects points of constant Doppler shift.*

Contours of constant line-of-sight velocity in a galaxy appear as a *velocity field*, as shown in Figure 7.4. Such a diagram is sometimes called a *spider diagram*.

The vertical line in the center represents the *minor kinematic axis*. The ovals near each end are contours of maximum positive and maximum negative projected rotation velocity and occur only if the true rotation curve is falling at large radii. A line drawn through them and the galaxy center represents the *major kinematic axis*. The major and minor kinematic axes are also called the *principal kinematic axes*. This velocity field shows only a few velocity contours, whereas a galaxy has a continuum of velocities from center to edge. Figure 7.5 shows an image made with a Fabry-Perot spectrometer, which records optical Doppler-shifted Hα emission from an object; the rotation curve (see Section 7.2) in this case is still rising at the edge.

Channel maps are useful for studying the distribution of H I throughout a galaxy. These maps are column density contours made for a particular velocity interval. In Figure 7.6, two channel maps are shown for an interacting galaxy pair, IC 2163 and NGC 2207. The higher-velocity channel traces emission from the small galaxy, whereas the lower-velocity channel traces emission from the large galaxy.

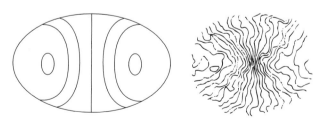

FIGURE 7.4   *(left) A spider diagram shows lines of constant velocity; the vertical line is the minor kinematic axis and has a velocity equal to the systemic velocity. (right) A spider diagram of M33. (Based on Bosma 1981 and Rogstad et al. 1976.)*

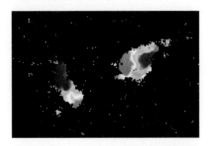

FIGURE 7.5   *Fabry-Perot image of the interacting galaxy pair Arp 240, showing velocity contours. The light gray sides on the upper part of the left-hand galaxy and the right part of the right-hand galaxy are blueshifted; the opposite sides are redshifted. (Image from the Electronic Universe website, G. Bothun, University of Oregon.)*

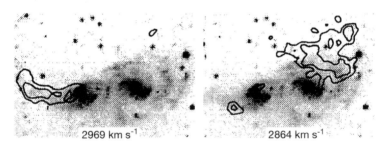

FIGURE 7.6   *Channel maps of IC 2163 (left-hand galaxy) and NGC 2207 (right-hand galaxy). The higher-velocity channel shows emission from the tail of IC 2163; the lower-velocity channel shows emission from the outer arms of NGC 2207. (From Elmegreen et al. 1995.)*

7.2 Rotation Curves

The rotation curve for a galaxy can be determined from the entire velocity field or by measuring only the velocities along the major axis if the motions are circular. On the basis of optical data, typical rotation curves in the 1960s resembled the one shown in Figure 7.7.

This curve shows two main components: The first is a linear rise from the center of the galaxy; the second is a steady decline in the outer parts. This decline would be expected if there were a large central mass that dominated the disk motions (such as satellite motion around Earth). This behavior is expected from Kepler's third law. For uniform circular motion, the circular velocity depends on the balance between the centrifugal force, $mv^2(r)/r$, and the gravitational force, $F_g = GMm/r^2$. For constant m, the velocity scales as $r^{-1/2}$.

Spectral line observations of H II regions and 21-cm aperture synthesis observations over the last two decades have resulted in more-sensitive measure-

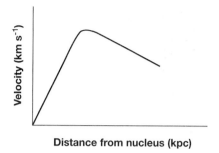

FIGURE 7.7   *A falling rotation curve, with (unrealistic) Keplerian rotation.*

ments than the early optical observations. Rotation curves are now observed to be flat or slowly rising rather than falling in the outer regions. A simplified rotation curve is shown schematically in Figure 7.8, and curves for nearby spirals based on 21-cm observations are shown in Figure 7.9.

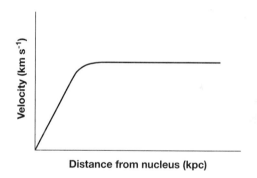

FIGURE 7.8   *A simplified rotation curve, which is flat in the outer parts.*

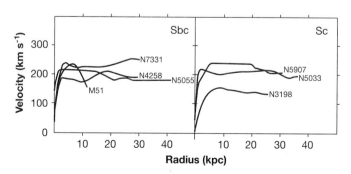

FIGURE 7.9   *H I rotation curves for some nearby spirals. (From Bosma 1981.)*

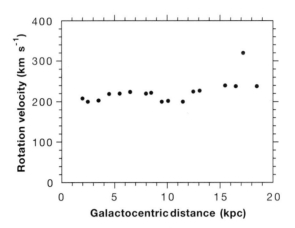

FIGURE 7.10   *Milky Way rotation curve, based on CO and H I observations.*
(Adapted from Fich and Tremaine 1991.)

For comparison, the rotation curve of the Milky Way beyond the bulge is
shown in Figure 7.10. It is based on observations of CO and H I emission, assuming that the Sun's distance R_o from the Galactic Center is 8.5 kpc and that
the Sun's rotational velocity is 220 km s^{-1}. The uncertainty in the distances out
to 13 kpc is about 0.5 kpc, but it increases to 4 kpc in the outer galaxy. The derived rotation velocities are accurate to 5 km s^{-1} inside 13 kpc, and 50 km s^{-1}
beyond that distance.

The flat part of a rotation curve in the outer part of a galaxy is evidence for
dark matter, because it shows that the total mass is still increasing with radius
even though there is little light. The linear part of the rotation curve is usually in
the bulge. The linear increase in velocity with radius means that the angular velocity, $\omega = v/r$, remains constant, so we call this *solid body rotation*. A plot of angular velocity with radius for the rotation curve in Figure 7.8 is flat in the inner regions and decreasing in the outer regions, as shown schematically in Figure 7.11.

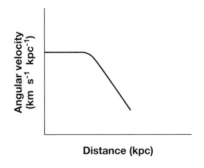

FIGURE 7.11   *A simplified angular velocity curve, which is flat for the solid*
body part and declines in the disk.

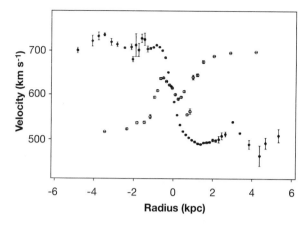

FIGURE 7.12 *Counter-rotating disk in the Sa galaxy NGC 3593. The open circles are from the stellar component; the dots are from the oppositely rotating ionized gas. (Based on [N II] and [S II] lines from Bertola et al. 1996.)*

The decrease in angular velocity with distance is known as *differential rotation* and is characteristic of large galaxy disks. Differential rotation means that two objects near each other but at different distances from the galaxy center gradually move away from each other. This separation occurs even if the two objects have the same linear speed (e.g., in a flat rotation curve galaxy), because the object farther from the galaxy center has a larger distance to cover in each revolution around the galaxy. The units for angular velocity are generally expressed in kilometers per second per kiloparsec (km s^{-1} kpc^{-1}). Note that an angular velocity is dimensionally the same as inverse time (that is, frequency).

Recent high-resolution observations of the centers of some galaxies indicate that *counter-rotating disks* can exist, as shown in Figure 7.12. These components are rotating opposite to the overall galaxy rotation and are most likely the result of a capture of material through an interaction or merger, as discussed further in Chapter 12.

7.3 Rotation Curve Fits

The peak rotation velocity v_{max} is a function of Hubble type and luminosity. For spiral galaxies of the same luminosity, early types have higher peaks than later types. There is a luminosity correlation as well, such that for a given Hubble type, more-luminous galaxies have higher peak velocities than do less-luminous galaxies. Figure 7.13 shows the range of rotation curves as a function of Hubble type and absolute magnitude (M_B), based on modeling optical observations by Vera Rubin and colleagues. The peak velocities for Sa galaxies range from about 100 km s^{-1} for an absolute blue magnitude of -18, up to 350 km s^{-1} for an

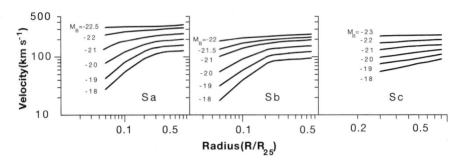

FIGURE 7.13   *Synthesized rotation curves for Sa, Sb, and Sc galaxies; absolute magnitudes (M_B) are indicated. (Adapted from data by Rubin et al. 1982, 1985.)*

absolute blue magnitude of -22.5. Similarly, Sb galaxies range from about 80 km s^{-1} to about 280 km s^{-1} for magnitudes -18 to -22.5, and Sc galaxies from 60 km s^{-1} to 220 km s^{-1} for magnitudes -18 to -23.

Galaxies reach their peak velocities at a galactocentric distance $R_{v_{max}}$ that depends on their size and Hubble type. A large spiral galaxy will usually reach its peak velocity at a distance that is smaller, compared with R_{25}, than the distance at which a small spiral galaxy of the same Hubble type will reach its peak velocity.

A correlation between peak velocity and the radius at which it occurs was recognized by de Vaucouleurs:

$$\log v_{max} + \log\left(\frac{2R_{v_{max}}}{D_{25}}\right) = 2.18 \pm 0.03$$

We can apply this relationship to the Milky Way. Using $v_{max} = 225$ km s^{-1} at $R_{v_{max}} = 7$ kpc, we find that D_{25} is at 23.4 kpc. This value is in good agreement with the 23.1 kpc value that de Vaucouleurs derived on the basis of photometric properties of our Galaxy.

Although the shape of a rotation curve is a function of both Hubble type and luminosity of a galaxy, the optical outer regions have been fit to a function of the luminosity alone by M. Persic and P. Salucci. This fit does not apply to the inner regions near the bulge. The slope, Δ, of the rotation curve for galactic radii between 1 and 3.2 disk scale lengths (where the outer limit is $\approx R_{25}$) is given by

$$\Delta = 0.12 + 0.096 \, (M_B + 21.5)$$

for absolute blue magnitude M_B. By definition, $\Delta = 0$ for a flat rotation curve. It is convenient to express the rotation curve as a *power law*, which is a good approximation for the outer optical regions:

$$v(R) \propto R^{\alpha}$$

The power α is related to the slope Δ by

$$\alpha = \frac{\log\left[\dfrac{2.2 + \Delta}{2.2 - 1.2\Delta}\right]}{\log 3.2}$$

$\alpha = 0$ for a flat rotation curve. In general, the power ranges from about -0.2 (falling curve) to $+0.2$ (rising curve). For the Milky Way, de Vaucouleurs estimated $M_B = -20.04$, which gives a slope of 0.256 and a power of 0.22.

This relationship between slope or power and galaxy luminosity makes it possible to approximate the shape of the rotation curve even for galaxies that are too difficult to measure, for example, face-on galaxies. The shape of the rotation curve is of fundamental importance in understanding spiral structure, as will be considered in Chapter 9.

As we saw in Chapter 6, the hydrogen disk generally extends beyond the optical disk. The *extended gradient*, EG, measures the slope of the rotation curve by comparing the velocities at $0.8\ R_{25}$ and $1.2\ R_{25}$ relative to the maximum velocity:

$$EG = \frac{[V_{1.2R_{25}} - V_{0.8R_{25}}]}{V_{max}}$$

On average, the slope is 0 (flat), but there is a range from -30% (falling rotation curve) to $+20\%$ (rising rotation curve). Late-type dwarf galaxies tend to have rotation curves with positive gradients. There is also a correlation with Arm Class (see Chapter 5); flocculent galaxies tend to have slightly positive slopes, and grand design galaxies tend to have slightly negative slopes.

The gradient of a rotation curve from 0.4 to $0.8\ R_{25}$ is called the *outer gradient*. It is correlated with environment in the sense that galaxies in dense clusters have negative outer gradients, whereas galaxies near the edges of clusters or in the field have 0 or positive outer gradients.

7.4 Tully-Fisher Relation

For galaxies with angular sizes that are too small to be resolved at 21 cm, all that can be measured is the total H I linewidth. The unresolved line profile has a characteristic *double-horn* shape, as shown in Figure 7.14. The line spread is due to the rotation of a galaxy, with blueshifted and redshifted motions along our line of sight. The "horns" occur because there is more gas with the same radial

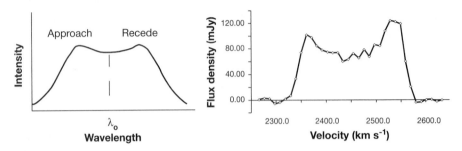

FIGURE 7.14 *(left) Double-horn H I profile of an unresolved galaxy, resulting from Doppler-shifted motions in a differentially rotating disk; λ_o is the wavelength corresponding to the systemic velocity. (right) H I profile of the edge-on spiral galaxy UGC 7170. (From Cox et al. 1996.)*

velocity in middisk than there is along a line of sight near the center of the galaxy. That is, the isovelocity region is larger in middisk, so the flux is larger.

There is an increase in the 21-cm linewidth of a galaxy with increasing absolute magnitude, as shown in Figure 7.15. This correlation is known as the *Tully-Fisher relation*. The correlation is best if infrared rather than blue magnitudes are used, because of smaller variations in dust extinction with galaxy inclination (see Chapter 4) at longer wavelengths. There appears to be a slightly different zero point in the relation for different Hubble types; types later than Sb have an offset of about -0.3 magnitude relative to earlier types (that is, for a given linewidth, later types are 0.3 mag brighter than earlier types). This result is related to the correlation between Hubble type and peak rotation velocity mentioned earlier. Low surface brightness galaxies have the same Tully-Fisher relation as normal spirals, but the scatter in the relation is greater for LSBs than for normal spirals.

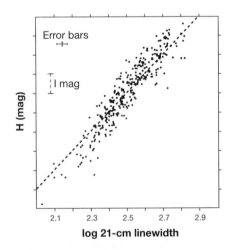

FIGURE 7.15 *Tully-Fisher relation using H band magnitudes. (From Aaronson et al. 1982.)*

The observed slope of the magnitude-log velocity plot was originally estimated to be approximately -10 (mag log km s^{-1}), which is equivalent to luminosity $L \propto v^4$. Recent observations indicate slopes closer to -7.7, with $L \propto v^{3.7}$ for spirals and irregulars, or $L \propto v^{2.5}$ for bright spirals alone.

The linewidth provides a measurement of a galaxy's absolute magnitude; that, plus the observed magnitude of the galaxy, may be used to derive the distance (see Chapter 3). This relation is particularly important for cosmology because distances are difficult to determine by other methods.

7.5 Mass Distribution in Disk Galaxies

Orbital periods of stars and gas can be used to deduce the mass distribution in a galaxy. Masses can be determined by comparing observed rotation curves with theoretical rotation curves based on a given mass distribution. A thin disk model for a galaxy gives a lower limit to the galaxy mass, whereas a spherical galaxy model gives an upper limit.

In a galaxy with spherical symmetry, the mass $M(r)$ interior to r is given by

$$M(r) = \int_0^r \rho(r) dV$$

for volume dV and density $\rho(r)$. We can determine $M(r)$ from the rotation curve $v(r)$ and the balance between centrifugal and gravitational forces,

$$M(r) \approx \frac{v(r)^2 r}{G}$$

The mass determined in this way is the dynamical mass M_{dyn}. This equation assumes that the velocity profile is due to rotation.

If highly flattened spheroids are used in models, the masses are 35% smaller. The mass determined from a thin disk approximation is

$$M(r) \approx \frac{2v(r)^2 r}{G\pi}$$

Note that $M(r)$ scales directly with radius for constant rotation velocity, which means that the mass continues to increase at large radii. Flat rotation curves in regions with no light indicate the presence of a *dark halo* of undetected matter. Dark halos can be 10 to 100 times more massive than the optical disk; this topic will be discussed further in Section 7.9.

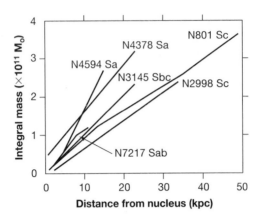

FIGURE 7.16   *Integrated mass as a function of radius for different Hubble types using a thin disk approximation. (Data from Rubin et al. 1978.)*

Dynamical masses of spiral galaxies are a function of Hubble type and galaxy size and range from $\sim 10^9$ $M_\odot$ to $\sim 10^{12}$ $M_\odot$. The masses of galaxies decrease on average from S0 to irregulars, because of decreasing size and slower maximum rotational velocity with later types. For the Milky Way, the total mass out to twice the Sun's distance of 8.5 kpc is 4.1×10^{11} $M_\odot$. Depending on the details of the outer rotation curve, recent mass estimates for the Milky Way reach as large as 5×10^{12} $M_\odot$ out to 60 kpc. Most of this mass is from the dark halo. Similarly, M31 has a mass of 3.4×10^{11} $M_\odot$ out to 10 kpc and $\sim 10^{12}$ $M_\odot$ out to 100 kpc for a flat rotation curve.

The density $\rho \propto M/r^3 \propto v^2 r/r^3$, so for a flat rotation curve (where v = constant), $\rho \propto 1/r^2$. The mass gradient, dM/dr, indicates how the mass is concentrated in a

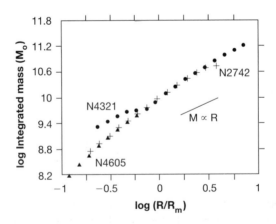

FIGURE 7.17   *Integrated mass in three Sc galaxies, scaled to a length R_m that contains 10^{10} $M_\odot$ within it. The solid line shows the slope if mass scales directly with radius. (Data from Burstein et al. 1982.)*

galaxy. Figure 7.16 shows the integrated mass versus the distance from the center of the galaxy for several Hubble types, based on observed rotation curves. Early-type spiral galaxies have steeper mass gradients than do late-type galaxies; however, large late-type galaxies have larger total masses than do small early types. For example, NGC 4594 has a steeper gradient than NGC 801 but less total mass because it is smaller.

When the radius is normalized to a length that contains a given amount of mass, the integral mass versus radius follows the same functional form in each galaxy. This relation is shown in Figure 7.17.

7.6 Early-Type Galaxy Rotations and Velocity Dispersions

Observations of motions in early-type galaxies are very difficult, because elliptical and S0 galaxies do not have much gas or many H II regions. This paucity of interstellar matter requires tedious measurements of weak stellar absorption lines.

The stars in elliptical galaxies orbit in three dimensions, with random orbital inclinations. The observed velocities are a combination of rotational velocities and velocity dispersions. Velocity curves can be obtained along different axes. The maximum velocity is much slower than in spirals: v_{max} is generally less than 100 km s^{-1}, with a mean value ~60 km s^{-1}, compared with Sa–Sc mean observed values of ~180 km s^{-1}. Lenticular galaxies have rotation properties similar to spirals, v_{max}~160–350 km s^{-1}, based on absorption line measurements of stars.

There is no Doppler shift along the kinematic minor axes in disk galaxies, since the motion there is transverse to our line of sight. In ellipticals, there are gradients along all directions, so no such preferred axis is measured; this result suggests that the ellipticity is not the result of flattening from rotation. The internal random motions are higher for ellipticals than for spirals: the velocity dispersion, or random motion, is σ_o~130–350 km s^{-1} in elliptical centers, compared with 75–215 km s^{-1} in early-type spirals. For spirals, $v_{max}/\sigma_o > 1$; for ellipticals, $v_{max}/\sigma_o < 1$ and is generally ≈ 0.2. This result implies that circular motions dominate in spirals, whereas random motions dominate in ellipticals. At any given flattening, elliptical galaxies rotate with a wide range of rates. The ratio $v_{max}/\sigma_o > 0.5$ in the bulges of spirals, which is consistent with models of oblate isotropic rotation. The smaller ratio for ellipticals suggests that triaxial (three different axis lengths) structures rather than oblate (two equal axes) structures are most common. More-luminous ellipticals have slower rotations than do less-luminous ones; faint ellipticals have rotations similar to spiral bulges.

Disky and boxy ellipticals (discussed in Chapter 5) differ in their kinematic properties. Disky ellipticals have a higher v_{max}/σ_o than do boxy ellipticals. This higher ratio may be due to the disklike part of the elliptical, with circular motions dominating.

7.7 The Fundamental Plane of Elliptical Galaxies

For elliptical galaxies, S. Faber and R. Jackson found that the luminosity scales approximately as $L \propto \sigma^4$ for velocity dispersion σ (which is measured approximately by the linewidth of the integrated stellar spectrum), as shown in Figure 7.18. Because absolute magnitude and diameter are related, a correlation also appears between diameter and σ, also shown in Figure 7.18. A normalized diameter D_n is commonly used; it is the diameter inside which the mean blue surface brightness Σ is 20.75 mag arcsec^{-2}. The diameter-velocity dispersion relation is sometimes called the $D_n - \sigma$ relation. These correlations are analogous to the Tully-Fisher relation and are also useful as distance indicators.

The correlation between elliptical galaxy luminosity and velocity dispersion has less scatter if the radius of the galaxy is also included, giving a best fit relation of

$$L \propto \sigma^{2.65} r_e^{0.65}$$

r_e is the observed radius enclosing half the light (see Chapter 5; it is related to D_n but defined differently). On the basis of elliptical and S0 galaxies, a similar fit is given by

$$\log r_e = 1.24 \log \sigma - 0.82 \log \Sigma_e$$

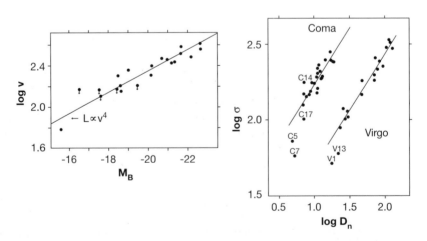

FIGURE 7.18 *(left) Line-of-sight velocity dispersion v versus absolute magnitude for elliptical galaxies. (From Faber and Jackson 1976.) (right) Normalized diameter D_n versus central velocity dispersion σ for cluster galaxies. (From Dressler et al. 1987.)*

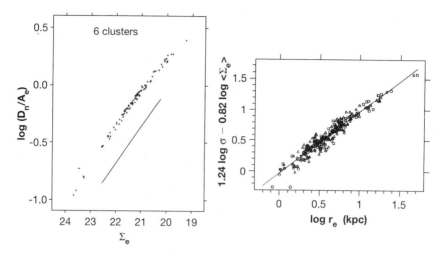

FIGURE 7.19 *The fundamental plane of elliptical galaxies: (left) The line has a slope of log D/A$_e$ ∝ Σ$_e$$^{4/5}$ for effective diameter A$_e$, surface brightness Σ$_e$, and diameter D$_n$. (Data from Dressler et al. 1987.) (right) The line has a slope of 1. (Data from Jørgensen et al. 1996.)*

for surface brightness Σ_e. This relation is shown in Figure 7.19. The *fundamental plane* is the name given to the two-dimensional space that galaxies occupy in the three-dimensional space of luminosity, velocity dispersion, and radius. The fact that galaxies are constrained in this way may reflect the processes by which they formed.

7.8 Masses of Early-Type Galaxies

Masses for E and S0 galaxies are more difficult to determine than are masses for spiral galaxies, because the internal velocities of elliptical galaxies have a larger component from random motions than from rotation. Thus, the mass determination is much more model-dependent for elliptical galaxies.

The *virial theorem* can be applied in order to estimate masses:

$$2T + U = 0$$

in equilibrium, for kinetic energy T and potential energy U. The kinetic energy in this case is given by

$$T = \frac{1}{2}M <v^2>$$

where M is the total mass of the galaxy; $<v^2>$ is the mass-weighted average of the square of the three-dimensional space velocities of stars relative to the center of mass of the galaxy and equals $3\sigma^2$. The velocity is usually estimated from the observed line-of-sight velocity dispersion σ in the nucleus, and σ^2 is assumed to be constant throughout the galaxy. The potential energy U of the stars is

$$U = -G\int_0^R \frac{M(r)dM}{r}$$

U is very uncertain. For spherical galaxies, the potential energy is $U = -0.33$ GM^2/r_e, where r_e is the isophotal radius that contains half the light; the constant changes for flattened galaxies. This isophotal radius also contains half the mass, assuming that the light distribution traces the mass. The potential energy can be derived by assuming that the M/L ratio is constant with radius in the optical part of the galaxy, and that the $r^{1/4}$ law for luminosity applies (although in dense environments the outer profiles of ellipticals sometimes depart from $r^{1/4}$ due to mass capture through mergers and stripping). The virial theorem then gives the mass

$$M = \frac{\sigma^2 r_e}{0.33G}$$

Elliptical galaxy masses that have been measured using the virial theorem range from ~3 × 10^{10} M_0 to ~3 × 10^{12} M_0.

The equation of hydrostatic equilibrium (see Chapter 3) can also be applied to deduce elliptical galaxy masses. For example, x-ray emission around ellipticals has a temperature distribution that depends on the local gravitational potential, which indicates a mass. The temperature is related to the pressure term through the perfect gas law. The results are largely model-dependent and require measurements of radial temperature gradients; masses determined in this way are a few times 10^{12} M_0. The kinematics and spatial densities of globular clusters in the outer parts of ellipticals similarly measure the galaxy mass.

7.9 Mass-to-Light Ratios and Dark Matter

The total mass M_{tot} is correlated with galaxy luminosity, as shown for S0 through Im galaxies in Figure 7.20. This relation also applies to dwarf and elliptical galaxies.

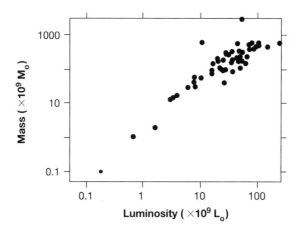

FIGURE 7.20   *Mass-luminosity relation for galaxies with types S0 through Im. (Based on Faber and Gallagher 1979.)*

The mass-to-light ratio, M/L, is a useful measure of how the mass (which is measured from the gravitational forces in a galaxy) compares with the luminosity (which is measured from starlight). The overall mass-to-light ratio in a galaxy is an indication of the proportions of dark matter and luminous matter; the startling conclusion is that a significant fraction of the matter in a galaxy must be dark. To estimate how much dark matter is needed to account for observations, let us review the mass-to-light ratios for stars.

As we saw in Chapter 4, stars on the main sequence have luminosities that scale with their mass: $L \propto M^{3.2}$ approximately. Table 7.1 emphasizes that the M/L ratio increases with later stellar types along the main sequence; masses have much less range than luminosities.

There is a correlation between M/L ratios and Hubble type, although there is a large spread in the data, as shown in Figure 7.21. The ratios are $M_{dyn}/L_B = 6.5$ $M_{\odot}/L_{\odot}$ for S0s, 5.1 for Sas, 4.3 for Sbs, 3.9 for Scs, and 3.7 for Sds. This similarity

TABLE 7.1   *M/L ratios for different stellar types*

Stellar type	Mass ($M_{\odot}$)	Luminosity ($L_{\odot}$)	M/L
O5 V	40	3×10^5	1.3×10^{-4}
A0 V	4	79	0.05
G2 V	1	1	1
M0 V	0.5	6.3×10^{-2}	7.9
M5 V	0.2	7.9×10^{-3}	25.3
F0 I	~20	~8×10^4	2.5×10^{-4}
F0 III	~5	~10^2	0.05
White dwarf	~1	~10^{-3}	~10^3

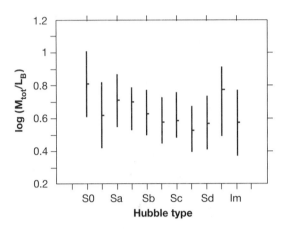

FIGURE 7.21   *Total mass-to-light ratio as a function of Hubble type. Dashes indicate median values; bars range from the 25th to the 75th percentile. (Based on Roberts and Haynes 1994.)*

between starlight M/L ratios and observed galaxy M/L ratios indicates that unseen matter does not strongly dominate the mass inside the Holmberg radius R_H. Dwarf spheroidal and dwarf irregular galaxies have average M/L ratios from ~5 to ~100 M_o/L_o; the ratio increases with decreasing galaxy luminosity.

The M/L ratio increases with increasing radius in a galaxy. Flat rotation curves with no corresponding starlight require dark matter; the fact that some galaxies have falling H I rotation curves indicates that the ratio of dark matter to luminous matter may vary in different galaxies. For flat rotation curves in spiral galaxies, the inferred M/L ratio is ~40 M_o/L_o near the optical edge, whereas gravitational effects in binary galaxies imply M/L ratios as high as 100 M_o/L_o in the tidal tails.

The M/L ratio for the solar neighborhood is estimated to be between ~1.5 and 5 M_o/L_o. The Ostriker-Caldwell model of the Milky Way, which assumes a rising rotation curve out to $1.6R_o$, sets R_H ~ 20 kpc and has 25% of the matter inside this radius in the form of dark matter. The Bahcall-Schmidt-Soneira fit to the rotation curve is a four-component model with a different density distribution for the bulge, disk, stellar halo, and dark matter halo. In the solar neighborhood, the density is 0.15 M_o pc^{-3}, of which 0.009 M_o pc^{-3} is dark matter.

For elliptical galaxies, the average mass/light ratio M/L_B is about 5–10 M_o/L_o, and may be as high as 20 M_o/L_o. Global anisotropic models for nearly spherical systems (E0–E2 galaxies) have been constructed that fit observed radial distributions well; these models assume a constant M/L ratio within the galaxy. It is very difficult to measure M/L as a function of radius for ellipticals. Observations of the radial intensity distributions indicate that, beyond the effective radius r_e, the M/L ratio increases with increasing radius in the same way as in spiral galaxies, as shown in Figure 7.22.

The match between models and elliptical galaxy photometric properties indicates that dark matter is minimal out to the effective radius r_e. The increasing

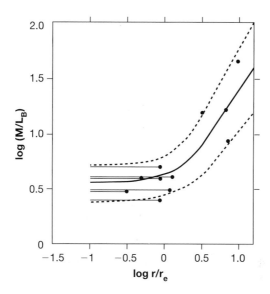

FIGURE 7.22 *The M/L ratio as a function of radius (scaled to the effective radius) is indicated for spirals (dots) and ellipticals (lines based on model, with a minimum and maximum limit shown by light dashed lines). The ellipticals and spirals have the same constant distribution in the inner parts and an increasing M/L in the outer parts, although there is a zero-point shift. (Based on Bertola et al. 1996.)*

M/L ratios in the outer parts of ellipticals suggest that the dark matter halo is probably substantial.

Galaxy colors can be used to determine a M/L ratio based on stellar population synthesis models. By converting from surface brightness to surface mass density using M/L for stars, we can predict a rotation velocity for a spiral galaxy disk of a given luminosity. The difference in velocity between the observed rotation curve and the predicted curve can be attributed to dark matter. A model for the dark halo density that is able to reproduce a flat rotation curve at large r is a spherical distribution of the form

$$\rho(r) = \frac{\rho_0}{\left(1 + \dfrac{r^2}{r_c^2}\right)}$$

where ρ_0 is the central density and r_c is the core radius, which must be fit to the data for a particular galaxy.

The corresponding dark mass within radius r is proportional to r for r >> r_c and dominates the mass at large radii. Luminous matter dominates the mass inside r_e, and both luminous and dark matter contribute substantially to the mass at intermediate radii. In dwarf galaxies, dark matter may dominate the mass even inside the optical radius. However, dwarf shapes are not well-known; dwarfs may be spherical, with velocity dispersions exceeding rotational velocities.

The mass of the disk compared with the mass of the halo has an empirical power law relation with luminosity for spiral galaxies:

$$\frac{M_{disk}(R_{25})}{M_{halo}(R_{25})} = \left(\frac{L_B}{L_B^*}\right)^{\gamma}$$

where $L_B^* = 10^{10} L_o$ (from the galaxy luminosity function; see Chapter 12), and the power γ is ~0.5. From this relation, it is evident that less-luminous galaxies have relatively larger halos. This result is also consistent with observations of dwarf galaxies, which are thought to have a higher proportion of dark matter than do spirals.

The composition of dark matter is unknown; some of it could be brown dwarfs, which are hypothetical low mass (<0.08 M_o) objects. They cannot achieve the 10^7 K temperature necessary for nucleosynthesis, but they might be detectable in the infrared. It is believed that clouds may fragment hierarchically and that low mass stars form more readily than high mass stars, so there could be very low mass brown dwarfs. However, if the initial mass function turns over at small masses, there would not be many low mass objects. Brown dwarfs may

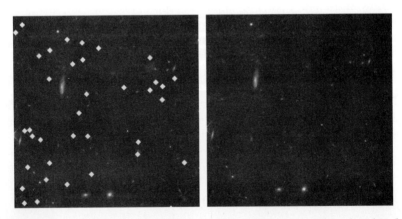

FIGURE 7.23   *A randomly selected area of the sky was imaged in the I band to search for faint halo stars in the Milky Way. (left) The white diamonds represent faint red stars that might have been expected on the basis of models. (right) Many distant galaxies are actually seen because the halo region in the Milky Way is so devoid of stars. (Imaged with Hubble Space Telescope WFPC2 by J. Bahcall, Institute for Advanced Study, Princeton; and NASA.)*

have been observed in the Hyades cluster. Figure 7.23 shows a Hubble Space Telescope image used to search for faint halo stars. The paucity of such stars suggests that they cannot account for all of the dark matter needed to match the rotation curve masses.

Dark matter could also be compact stellar remnants; it is difficult to detect most white dwarfs, neutron stars, and black holes. Recent speculations include the possibility that dark matter may be very cold molecular gas. We can put limits on how much mass can be accounted for by ordinary matter, as we will explain below; the results of models indicate that most dark matter must be something other than ordinary matter.

Matter consists of leptons and hadrons. Leptons are elementary particles that have half-integer spin and do not interact via the strong force; they include electrons. Hadrons are particles that interact by the strong force and have spins that are multiples of one-half; they include *baryons* and *mesons*. Baryons include protons and neutrons, which consist of 3 *quarks* each. Quarks are fundamental elementary particles and come in 18 different types. Cosmological nucleosynthesis predicts the abundances of elements relative to hydrogen; we can sum these to estimate the present mean density of all baryons, ρ_B.

A *critical density* ρ_c is the density needed to close the Universe: that is, ultimately to halt expansion. The ratio of the present density ρ_0 to ρ_c is denoted by the symbol Ω. $\Omega < 1$ corresponds to a Universe that has less than the critical density and will expand forever; $\Omega > 1$ is a Universe that will eventually collapse. The value of ρ_c depends on H_0^2, where H_0 is the Hubble constant that represents the expansion rate of the Universe:

$$\rho_c = \frac{3H_0^2}{8\pi G}$$

For $H_0 = 75$ km s^{-1} Mpc^{-1}, $\rho_c = 1.1 \times 10^{-29}$ g cm^{-3}. The present density ρ_0 can be estimated through several different measurements, including those based on motions of clusters of galaxies and on rotation curves of galaxies. The value of Ω is very uncertain but is probably ~0.1 to ~1.

We can compare the baryon density ρ_B with the critical density ρ_c to get a value $\Omega_B = \rho_B/\rho_c$. Star luminosities and gas emission lead to a ratio of $[\rho_{stars+gas}/\rho_c]$ ~ 0.01. If Ω is between 0.1 and 1, then between 10 and 100 times more mass is needed than that of the visible stars and gas. Nucleosynthesis models match observed chemical abundances only if $0.02 < \Omega_B(H_0/75$ km s^{-1}Mpc$^{-1})^2 < 0.03$. This means that there could be two or three times the mass of visible stars and gas in the form of undetected baryonic matter, but this is not enough to match the observed Ω.

Many types of nonbaryonic particles have been postulated to account for the "missing mass." There are *cold dark matter* (CDM) models of particles that are massive and therefore slow-moving; as a result, they are bound to galaxies. Some cold dark matter candidates include *MACHOs*, an acronym for massive astrophysical condensed halo objects. Some stars in the Large Magellanic Cloud show variability that may be caused by foreground MACHOs in the Milky Way halo with masses 0.03–0.3 M_o. This variability of stars behind dark objects is caused by gravitational *microlensing*.

Another possibility for nonbaryonic dark matter is weakly interacting massive stable elementary particles, dubbed *WIMPs*, which supposedly formed in the hot dense early Universe. Several plausible hypothetical candidates include *photinos*, which are partners of photons; *zinos*, partners of Z particles, and *winos*, partners of W particles. W and Z particles are intermediate bosons (particles with integer spin) and have nearly 100 times the mass of a proton; W particles are charged, and Z particles are neutral. Zinos and winos have expected masses equivalent to about 10 protons.

Other candidates for dark matter include *axions* and *neutrinos*; the neutrino is the only one known to exist, but its mass is unknown. Neutrinos are moving close to the speed of light and therefore are not bound to galaxies, so they would not account for the missing mass in galaxies. They are part of *hot dark matter* models. Cosmological implications for these models will be discussed further in Chapter 12.

Exercises

1. Determine the power law fits to the rotation curves for NGC 4321, NGC 5055, and NGC 5194, using only their absolute blue magnitudes.
2. Using the de Vaucouleurs absolute blue magnitude for the Milky Way, determine our galaxy's mass out to R = 8.5 kpc. Assume a power law fit for the rotation curve, using v = $10^5 R^\alpha$, for v in km s^{-1} and R in kpc.
3. Assume that all stars obey the mass-luminosity relationship L = $M^{3.2}$ for L and M in solar units. Determine the mass-to-light ratio (i.e., M/L) for the cases of:
   a. 1000 M stars and 10 O stars
   b. 1000 M stars and 100 G stars
   c. 10000 M stars, 100 G stars, and 1 O star
   d. Comment on these results in terms of the M/L ratio for the solar neighborhood and for flat rotation curves.

Unsolved Problems

1. What is dark matter?
2. What are the initial conditions that determine how fast a galaxy rotates?

Useful Websites

Fabry-Perot image observations showing galaxy rotation are available through the Electronic Workshop at http://zebu.uoregon.edu

Further Reading

Aaronson, M., R. Tully, J. Fisher, B. Siegman, J. Huchra, J. Mould, H. van Woerden, W. Goss, P. Chamaraux, and U. Mebold. 1982. A catalog of infrared magnitudes and H I velocity widths for nearby galaxies. *Astrophysical Journal* 50:241.

Ashman, K. 1992. Dark matter in galaxies. *Publications of the Astronomical Society of the Pacific* 104:1109.

Bahcall, J., M. Schmidt, and R. Soneira. 1982. On the interpretation of rotation curves measured at large galactocentric distances. *Astrophysical Journal Letters* 258:L23.

Bernstein, G., P. Guhathakurta, S. Raychaudhury, R. Giovanelli, M. Haynes, T. Herter, and N. Vogt. 1994. Tests of the Tully-Fisher relation. I. Scatter in infrared magnitude versus 21-cm width. *Astronomical Journal* 107:1962.

Bertola, F., P. Cinzano, E. Corsini, A. Pizzella, M. Persic, and P. Salucci. 1993. Counterrotating stellar disks in early-type spirals: NGC 3593. *Astrophysical Journal Letters* 458:L67.

Bertola, F., A. Pizzella, M. Persic, and P. Salucci. 1996. Dark matter halos in elliptical galaxies. *Astrophysical Journal Letters* 416:L45.

Binney, J. 1982. Dynamics of elliptical galaxies and other spheroidal components. *Annual Review of Astronomy and Astrophysics* 20:399.

Binney, J. 1992. Warps. *Annual Review of Astronomy and Astrophysics* 30:51.

Bosma, A. 1981. 21-cm line studies of spiral galaxies. I—Observations of the galaxies NGC 5033, 3198, 5055, 2841, and 7331. II—The distribution and kinematics of neutral hydrogen in spiral galaxies of various morphological types. *Astronomical Journal* 86:1791.

Burstein, D., V. Rubin, N. Thonnard, and K. Ford. 1982. The distribution of mass in Sc galaxies. *Astrophysical Journal* 253:70.

Carr, B. 1994. Baryonic dark matter. *Annual Review of Astronomy and Astrophysics* 32:531.

Casertano, S., and J. van Gorkom. 1991. Declining rotation curves—the end of a conspiracy? *Astronomical Journal* 101:1231.

Cox, A., L. Sparke, G. van Moorsel, and M. Shaw. 1996. Optical and 21-cm observations of the warped, edge-on galaxy UGC 7170. *Astronomical Journal* 111:1505.

Dressler, A., D. Lynden-Bell, D. Burstein, R. Davies, S. Faber, and R. Terlevich. 1987. Spectroscopy and photometry of elliptical galaxies. I. A new distance estimator. *Astrophysical Journal* 313:42.

Elmegreen, B. G., D. M. Elmegreen, and L. Montenegro. 1992. Optical tracers of spiral wave resonances in galaxies. II. Hidden three-arm spirals in a sample of 18 galaxies. *Astrophysical Journal (Suppl.)* 79:37.

Elmegreen, D., M. Kaufman, E. Brinks, B. Elmegreen, and M. Sundin. 1995. The interaction between spiral galaxies IC 2163 and NGC 2207. I. Observations. *Astrophysical Journal* 453:100.

Faber, S., and J. Gallagher. 1979. Masses and mass-to-light ratios of galaxies. *Annual Review of Astronomy and Astrophysics* 17:135.

Faber, S., and R. Jackson. 1976. Velocity dispersions and mass-to-light ratios for elliptical galaxies. *Astrophysical Journal* 204:668.

Fich, M., and S. Tremaine. 1991. The mass of the galaxy. *Annual Review of Astronomy and Astrophysics* 29:409.

Giovanelli, R., M. Haynes, L. DaCosta, W. Freudling, J. Salzer, and G. Wegner. 1997. The Tully-Fisher relation and H_o. *Astrophysical Journal Letters* 477:1.

Jørgensen, I., M. Franx, and P. Kjærgaard. 1996. The fundamental plane for cluster E and S0 galaxies. *Monthly Notices of the Royal Astronomical Society* 280:167.

Kent, S. 1990. Dark matter in elliptical galaxies. In *Astronomical Society of the Pacific Conference Series 10: Evolution of the Universe*, ed. R. Kron, 109. San Francisco: Astronomical Society of the Pacific.

Kormendy, J. 1990. Scaling laws for dark matter in late-type galaxies. *Astronomical Society of the Pacific Conference Series 10: Evolution of the Universe*, ed. R. Kron, 33. San Francisco: Astronomical Society of the Pacific.

Ostriker, J., and J. Caldwell. 1983. A model for the galaxy with rising rotational velocity. In *Kinematics, dynamics, and structure of the Milky Way: Proceedings of the workshop on the Milky Way*. Dordrecht: D. Reidel Publishing.

Persic, M., and P. Salucci. 1990. Dark matter in spiral galaxies. *Astrophysical Journal* 355:44.

Persic, M., and P. Salucci. 1995. Rotation curves of 967 spiral galaxies. *Astrophysical Journal (Suppl.)* 99:501.

Prugniel, P., and E. Simien. 1996. The fundamental plane of early-type galaxies: Stellar populations and mass-to-light ratio. *Astronomy and Astrophysics* 309:749.

Roberts, M., and M. Haynes. 1994. Physical parameters along the Hubble sequence. *Annual Review of Astronomy and Astrophysics* 32:115.

Rogstad, D.H., M.C.H. Wright, and I.A. Lockhart. 1976. Aperture synthesis of neutral hydrogen in the galaxy M33. *Astrophysical Journal* 204:703.

Rubin, V., D. Burstein, K. Ford, and N. Thonnard. 1985. Rotation velocities of 16 Sa galaxies and a comparison of Sa, Sb, and Sc rotation properties. *Astrophysical Journal* 289:81.

Rubin, V., W. Ford, and N. Thonnard. 1978. Extended rotation curves of high-luminosity spiral galaxies. IV. Systematic dynamical properties, Sa–Sc. *Astrophysical Journal Letters* 225:L107.

Rubin, V., N. Thonnard, K. Ford, and D. Burstein. 1982. Rotational properties of 23 Sb galaxies. *Astrophysical Journal* 261:439.

Solanes, J., R. Giovanelli, and M. Haynes. 1996. The H I content of spirals. I. Field galaxy H I mass functions and H I mass-optical size regressions. *Astrophysical Journal* 461:609.

Sprayberry, D., G. Bernstein, C. Impey, and G. Bothun. 1995. The mass-to-light ratios of low surface brightness spiral galaxies: Clues from the Tully-Fisher relation. *Astrophysical Journal* 438:72.

Tremaine, S. 1992. The dynamical evidence for dark matter. *Physics Today* 45:28.

Tully, R.B., and J.R. Fisher. 1977. A new method of determining distances to galaxies. *Astronomy and Astrophysics* 54:661.

Whitmore, B., D. Forbes, and V. Rubin. 1988. Rotation curves for spiral galaxies in clusters. II. Variations as a function of cluster position. *Astrophysical Journal* 333:542.

CHAPTER 8

Milky Way Kinematics and Structure

Chapter Objectives: to understand the distribution and motions of objects in our Galaxy

---

Toolbox:

populations	Oort's A and B
scale height	terminal velocities

8.1 Stellar Populations and Metallicities

In the 1940s, Walter Baade observed M31 and noticed that the brightest stars in the bulge and in the halo have redder colors than those in the spiral arms. The differences in color-magnitude diagrams and space distributions led Baade to the idea of two distinct groups of stars, which he called *Population I* and *Population II*. Astronomers recognize that there is a continuous range in the properties of stars within the two populations, but it is helpful to identify the key components of the main divisions.

Population I objects are young, blue, and metal-rich; they are concentrated in the plane and are associated with gas and dust. They have nearly circular orbits around the Galactic Center. Population II stars are old and red. They show very little or no systemic rotation as a group but instead have highly eccentric orbits over large radii. The halo stars are metal-poor and exist in a nearly spherical halo, or spheroid, surrounding the Galaxy; the bulge stars are less metal-poor but are also spherically distributed. Because even halo Population II stars have more metals than would have been produced in the early Universe, many astronomers believe that there was a previous generation of stars, too. Searches for these hypothetical *Population III* stars as the earliest, nearly zero-metal stars have been inconclusive. Table 8.1 lists properties of the main components of the populations.

Based on observations that metallicity decreases radially outward in the Galaxy, a scenario for our Galaxy's evolution has emerged in which the halo component

TABLE 8.1 *Milky Way properties*

Location	Population	Age (10⁹ yr)	z scale height (pc)	Metals Z/Z₀	Velocity dispersion (km s⁻¹)
Spiral arm	Population I: gas, dust, open clusters, associations, OB V, supergiants, Cepheids type I, T Tauri's	<0.1	120	1–2	15
Young disk	Population I: A-F V, A-K III	~ 1	200	1–2	25
Intermediate disk	Population I: Sun, GKM V, PN	~ 5	400	0.5–1	50
Old disk	Population I: K-M V, subgiants (II), red giants (III), long-period variables, RR Lyraes	<10	700	0.2–0.5	80
Bulge	Population II: globular clusters, RR Lyraes	10–15	>2000	10^{-3}–0.03	135–225
Halo	Population II: globular clusters, RR Lyraes	10–15	>2000	≥ 1	135–225

Source: Adapted from Mihalas and Binney 1981, tables 4–18 and ch. 4.

formed as a spherical and radially contracting system. Stars formed throughout the contraction of the halo's gas, so stars that formed first have larger orbital radii than do stars that formed later. There was a progressive metallicity enrichment of the halo gas as the massive stars evolved. Therefore, metallicity increased with contraction. There is currently a large (as much as a factor of 10) dispersion in metallicity at any given position in the inner Galaxy, due to mixing of old and young stars. The average metallicity and dispersion in metallicity decreases with increasing radius. In Figure 8.1, stars that formed earlier in the stellar halo are represented by more extended orbits than those that formed later. Inside volume C, there is a mix of star ages and therefore metallicities, whereas out to A there are older stars with smaller metallicities and a smaller spread in metallicities.

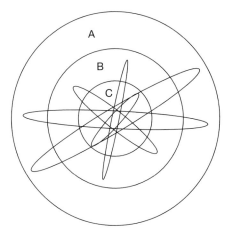

FIGURE 8.1 *Population II objects travel on highly eccentric orbits through the disk. The average metallicity in volume C is greater than that in B or A.*

Both the metal-poor halo and the less metal-poor bulge include many globular clusters and RR Lyrae stars; the spheroidal component contains virtually no gas or dust today, except for the Galactic Center. There are faint Population II subdwarfs in the spheroidal component that can only be detected locally. They have ages of 10^{10}–1.5×10^{10} years, with masses <0.8 M_o (more massive spheroidal component stars are already in the white dwarf stage). The brightest stars in the spheroidal component are red giants, whose maximum absolute magnitudes are -2.7.

The disk is believed to be about 10^{10} years old, on the basis of color-magnitude diagrams of clusters. This age is substantially younger than the halo, so the disk presumably formed out of metal-enriched material from halo evolution. The disk metallicity increased on average by a factor of 3 to 5 over this 10^{10}-yr period because of continued star formation and evolution, although the rate varied radially.

8.2 Stellar Distributions Perpendicular to the Plane

As the disk formed, cloud-cloud collisions may have circularized the gas motions. This idea is consistent with the observation that all Population I stars have approximately circular orbits about the galaxy. Older, more metal-poor disk stars have further excursions from the midplane than younger stars, so the disk thickness and metallicity are functions of the age. Theoretical analyses by Lyman Spitzer and Martin Schwarzschild suggested that the variation of scale height with spectral type is a consequence of the increased number of dynamical interactions that older stars have had with giant molecular clouds. Each perturbation can contribute to the random velocity of a star, so there are greater excursions among older stars.

Population I stars are concentrated toward the Galactic plane, with a decrease in number density, D(z), perpendicular to the plane. Close to the plane, the distribution is Gaussian; farther from it, the distribution is better fit by an exponential:

$$D(z) = D_o e^{-z/\beta}$$

where z is the distance from the midplane, β is the vertical scale height, and D_o is the midplane density for a particular spectral type. The *surface density*, or number of stars per square parsec, is given by the integral of the number density over the disk thickness, which is equivalent to the product of twice the scale height and the midplane density. This is a quantity that can be measured in other galaxies.

Early-type stars have smaller scale heights than late-type stars, as first recognized by B. Lindblad. The distribution with type is shown in Figure 8.2.

In addition to this distribution of stars, sometimes called the *thin disk*, there is a *thick disk* based on the measurements of K dwarf and subdwarf stars. Its distribution is not well established; the vertical density has an approximately exponential decrease with increasing height above the plane, with a scale height of about 1000 pc. The radial exponential scale length is similar to that of the thin disk, ~4 ±1 kpc. The mean metallicity and color-magnitude diagram of the thick disk resemble those of metal-rich globular clusters.

Based on star counts at high galactic latitudes between galactocentric distances of 4 and 12 kpc, the stellar spheroidal component of the Galaxy has been modeled to have a de Vaucouleurs radial profile (see Chapter 5).

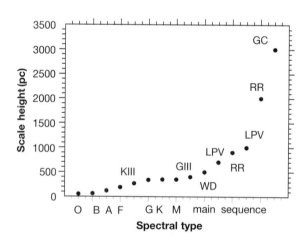

FIGURE 8.2 *Variation of vertical scale height with stellar type. GC, globular clusters; LPV, long-period variable; RR, RR Lyraes; WD, white dwarf. (Based on data from Mihalas and Binney 1981.)*

The scale heights for the gas components are similar to those for OB stars; the CO half-thickness is ~60 pc from the midplane of the galaxy, compared with the H I half-thickness of ~130 pc.

8.3 Standards of Rest

The center of mass of our Galaxy defines the *Fundamental Standard of Rest*. The *Local Standard of Rest*, or LSR, is the uniform circular motion of an object at a given distance from the Galactic Center. Stars and gas clouds acquire random motions through interactions with other disk material, so their circular orbits become perturbed. These perturbations can be in the radial or azimuthal direction or perpendicular to the disk. The individual components of a star's motion relative to its LSR are known as *peculiar velocities*. The total motion of a star in the radial direction is known as the Π *component*; in the azimuthal direction, the θ *component*, and in the perpendicular direction, the Z *component*. The Π and Z components of the LSR are 0 because that reference frame assumes uniform circular motion; the azimuthal component of the LSR velocity is θ_0. The deviation of the star or cloud components of motion from the LSR components are known as the $u, v,$ and w *components* of peculiar motion, defined as:

$$u \equiv \Pi - \Pi_{LSR} = \Pi$$
$$v \equiv \theta - \theta_{LSR} = \theta - \theta_0$$
$$w \equiv Z - Z_{LSR} = Z$$

The goals of stellar kinematics are to understand the LSR motions as a function of radius, to understand the motions of particular types of stars with respect to the LSR, and to understand the Sun's motion with respect to the LSR.

8.4 Solar Motion and Velocity Dispersions

We can measure a star's motion with respect to the Sun, which we designate by:

$$U = u - u_o = \Pi - \Pi_o$$
$$V = v - v_o = \theta - \theta_o$$
$$W = w - w_o = Z - Z_o$$

where the velocity components with subscript o represent the Sun. The Sun's components are determined with respect to a large number of stars with a common stellar type. Each star of a particular type will have random radial, azimuthal, and vertical motions, but the average for that type will have approximately uniform circular motion in the plane of the disk.

The average results for the Sun's peculiar velocities, determined by Baade and Mayall, are approximately:

$$u_o = -9 \text{ km s}^{-1}$$
$$v_o = 12 \text{ km s}^{-1}$$
$$w_o = 7 \text{ km s}^{-1}$$

The Sun's peculiar speed is given by the square root of the sum of the squares of the components, $(u_o{}^2 + v_o{}^2 + w_o{}^2)^{1/2} = 16.5$ km s^{-1}. The direction in which the Sun is moving is known as the *apex of solar motion*; it is inward toward the Galactic Center and upward out of the plane relative to the LSR, in the direction $l = 53°, b = 25°$.

For a particular stellar type, each star's velocity varies around the average with an approximately Gaussian distribution. That is, the number of stars, n, with a peculiar velocity component between u and u+du is given by

$$n(u) \, du \propto \exp\left[-\frac{u^2}{2<u>^2} \right] du$$

This distribution is shown in Figure 8.3.

Later spectral types have a broader spread of peculiar velocities than earlier spectral types. This result is consistent with the different scale heights mentioned earlier: The older stars have had more time than the younger stars to acquire random motions due to interactions with disk material.

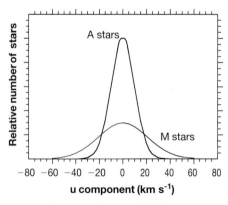

FIGURE 8.3   *Velocity dispersions for early- and late-type stars.*

Now suppose that we want to determine the number of stars that have particular peculiar velocities u, v, and w in the range (u, u+du), (v, v+dv), (w, w+dw). That number is an extension in three dimensions of the curves in Figure 8.3. In particular, the number n of stars per unit volume with velocities in a unit interval can be represented by:

$$n(u, v, w) \, du \, dv \, dw \propto \exp\left[-\left(\frac{u^2}{2<u>^2} + \frac{v^2}{2<v>^2} + \frac{w^2}{2<w>^2}\right)\right] du \, dv \, dw$$

This distribution function is known as the *velocity ellipsoid*, because the exponent has the same form as the expression for an ellipsoid,

$$\frac{x^2}{a^2} + \frac{y^2}{b^2} + \frac{z^2}{c^2} = 1$$

The principal axis of the velocity ellipsoid, called the *longitude of the vertex*, for stars in the vicinity of the Sun points approximately toward the Galactic Center. This result is a consequence of stars being in near-circular orbit about the Galactic Center, as proposed by K. Schwarzschild. Some stars show a *vertex deviation*, which reflects motions that may result from peculiar velocities of local star-forming regions. In particular, the young stars associated with *Gould's Belt* have a large vertex deviation and an asymmetric spatial distribution that is inclined 16° with respect to the plane of the Galaxy; see Figure 8.4. These stars are evidently part of an expanding group whose motions reflect the local velocity field of the spiral arm in which they formed. Older stars had time to come into

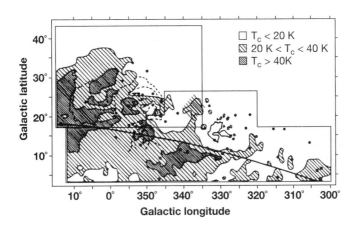

FIGURE 8.4 *Gould's Belt in H I. The arc represents the equator of the belt. (From Pöppel 1997.)*

equilibrium with the overall galactic potential, so their vertex deviations are smaller.

The total velocity dispersion, σ, of a group of stars may be represented by

$$\sigma = (\ <u^2> \ + \ <v^2> \ + \ <w^2> \ )^{1/2}$$

For all Population I stars, the measured velocity dispersions are such that $<u^2>$ is greater than $<v^2>$, which is greater than $<w^2>$. The ratio of the v component to the u component depends on the disk rotation properties through a star's epicyclic motion (see Chapter 9). For A0 stars, the peculiar velocity components are typically u = 15 km s^{-1}, v = 9 km s^{-1}, and w = 6 km s^{-1}; for K0 stars, u = 28 km s^{-1}, v = 16 km s^{-1}, and w = 11 km s^{-1}. In general, OB stars have approximately equal u and v components; other spectral types have the rms value of the v component, about two-thirds that of the u component. The kinematic properties of stars change abruptly for types later than F5, a result known as *Parenago's discontinuity*. Types later than F5 have velocity dispersions about twice that of earlier types. The break in the main sequence occurs for stars having an age approximately equal to the disk age.

Spheroidal component stars, with velocity dispersions greater than 65 km s^{-1}, behave very differently kinematically than low-velocity (i.e., Population I) stars. High-velocity stars have a low systemic rotation rate, so they lag with respect to the LSR. Most high-velocity stars in the solar neighborhood have a *perigalacticon* (closest approach to the Galactic Center) of about 4 kpc and an *apogalacticon* (farthest distance from the Galactic Center) of about 15 kpc, although some have an apogalacticon as much as 40 kpc. Some Population II stars have retrograde orbits because their peculiar velocities exceed the spheroidal component's small rotational velocity and are in the negative direction. In general, the dominant peculiar velocity component of spheroidal component stars is in the perpendicular direction, $<w^2>^{1/2} \sim$ 70–90 km s^{-1}.

8.5 Oort's Constants

We would like to be able to determine the motion of a star with respect to the Galactic Center in order to understand the overall kinematics of the Galaxy. We begin by measuring a star's velocity with respect to the Sun, and we use what we have learned about the Sun's motions and the LSR to translate this velocity into a galactic motion. The relative radial velocity, V_R, is given by the component of the star's circular velocity along our line of sight, $\theta \cos(\alpha)$, minus the Sun's component of motion along that line of sight, $\theta_o \sin(l)$, for galactic longitude l and angle α between the line of sight to the star and its space velocity vector (see Figure 8.5); that is,

$$V_R = \theta \cos(\alpha) - \theta_o \sin(l)$$

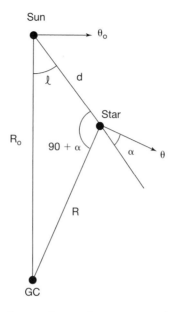

FIGURE 8.5 *The Sun's circular motion and a star's circular motion about the Galactic Center (GC).*

Using the law of sines,

$$\frac{\sin(l)}{R} = \frac{\sin(90° + \alpha)}{R_o} = \frac{\cos(\alpha)}{R_o}$$

then the star's radial velocity can be rewritten as

$$V_R = \frac{\theta}{R} R_o \sin(l) - \theta_o \sin(l)$$

or

$$\boxed{V_R = R_o(\omega - \omega_o) \sin(l)}$$

where $\omega_o = \theta_o/R_o$ is the angular velocity of the Sun and $\omega = \theta/R$ is the angular velocity of the star. This equation is a general expression for radial velocity and is valid everywhere in the disk as long as uniform circular motion applies.

Note that the distance R of an object from the Galactic Center can be determined from geometry once its distance from the Sun is known:

$$R^2 = R_o^2 + d^2 - 2R_o \, d \, \cos(l)$$

as evident from Figure 8.5.

For distances d between the Sun and a star that are small compared with the distance to the Galactic Center, the measured radial velocity of a star can be related to the angular velocity of the Sun and its velocity gradient, $(d\theta/dR)_{R_o}$, with a first-order Taylor series expansion (which is an approximation for the value of a function in the vicinity of a point, based on derivatives at that point), which gives

$$\omega - \omega_o \approx \left(\frac{d\omega}{dR}\right)_{R_o} (R - R_o)$$

The derivative is given by

$$\left(\frac{d\omega}{dR}\right)_{R_o} = \frac{1}{R_o}\left(\frac{d\theta}{dR}\right)_{R_o} - \frac{\theta_o}{R_o^2}$$

so that, to first order, we have

$$V_R = \left[\left(\frac{d\theta}{dR}\right)_{R_o} - \frac{\theta_o}{R_o}\right](R - R_o) \sin(l)$$

For distances d that are small compared with the Sun's distance from the Galactic Center, we have $R_o - R \approx d \, \cos(l)$, so the radial velocity can be rewritten as

$$V_R \approx \left[\frac{\theta_o}{R_o} - \left(\frac{d\theta}{dR}\right)_{R_o}\right] d \, \sin(l) \, \cos(l)$$

From trigonometry, $\sin(l) \cos(l) = 1/2 \, \sin(2l)$. Now define a quantity that incorporates the terms in brackets:

$$A = \frac{1}{2}\left[\frac{\theta_o}{R_o} - \left(\frac{d\theta}{dR}\right)_{R_o}\right]$$

This quantity is known as *Oort's constant A*, which is an angular velocity that is a measure of shear in a galaxy. Shear tells how an extended object is stretched out as it experiences differential rotation of the disk.

We can use these relations to rewrite the observed radial velocity as

$$V_R = A\, d \sin(2l)$$

which is good locally out to a few kiloparsecs. This is a useful expression, since the distance to an object is a function of local motions (incorporated in A) and the galactic longitude at which it is observed. No other assumptions about galactic rotation need to be made, except that the disk has uniform circular motion. This expression directly shows that there is a double sine wave relation between velocity and longitude for local objects.

Oort's A is measured by averaging over several different types of objects, such as Cepheids or A stars, in order to try to avoid widespread peculiar velocities. Its value is 14.5 ± 1.5 km s^{-1} kpc^{-1}.

Another constant, *Oort's constant B*, is defined as

$$B = -\frac{1}{2}\left[\frac{\theta_o}{R_o} + \left(\frac{d\theta}{dR}\right)_{R_o}\right]$$

This constant is equivalent to the quantity $(A - \omega_o)$ and has the measured value of -12 ± 3 km s^{-1} kpc^{-1} based on observations of the proper motions (i.e., motions transverse to our line of sight) of several groups of stars; it involves tangential velocities. In a derivation analogous to that for the radial velocity, the tangential velocity is given by

$$V_T = \theta \sin(\alpha) - \theta_o \cos(l)$$

and since

$$R \sin(\alpha) = R_o \cos(l) - d$$

we have

$$V_T = \frac{\theta}{R}[R_o \cos(l) - d] - \theta_o \cos(l)$$

$$= (\omega - \omega_o)R_o \cos(l) - \omega\, d$$

which is valid for strictly circular rotation. Then from Oort's constants, we can rewrite the tangential velocity as

$$V_T = d[A \cos(2l) + B]$$

From the definition of Oort's constants, we see that the Sun's angular velocity can be expressed as

$$\omega_o = \frac{\theta_o}{R_o} = A - B$$

and the local derivative of the circular velocity as

$$\left(\frac{d\theta}{dR}\right)_{R_o} = -(A + B)$$

We can relate the velocity dispersions to Oort's constants also, which is useful for obtaining a value for B:

$$\frac{<u>}{<v>} = \frac{-B}{A - B}$$

There are several different fits to the values of R_o and θ_o, based on observations of CO, planetary nebulae, H II regions, Cepheids, and H I emission. The standard values adopted by the International Astronomical Union (IAU) in 1985 are 8.5 kpc and 220 km s^{-1}, respectively, although models ranging from 8.2 to 10.0 kpc and 216 km s^{-1} to 250 km s^{-1} give consistent fits with the data.

Note that the Sun's orbital period around the galaxy, P_{rev}, is given by

$$P_{rev} = \frac{2\pi R_o}{\theta_o} = \frac{2\pi}{(A - B)}$$

$$= 0.24 \text{ kpc km}^{-1} \text{ s} = 7.4 \times 10^{15} \text{ s} = 2.4 \times 10^8 \text{ yr}$$

8.6 Radial Velocities

The distribution of gas in the Milky Way is determined by observing its Doppler-shifted emission as a function of direction. Typically three to five H I clouds are along a given kiloparsec line of sight, but the peak emission from each cloud is at a distinct velocity. We can make use of kinematic information about the disk to infer the distance to a gas cloud. That is, if we know the rotation curve of the Galaxy and assume circular rotation, we can calculate the expected distance from the Doppler-shifted velocity of an object at a given longitude.

The polar coordinate system for our Galaxy is centered on the Sun. Galactic longitude increases in the counterclockwise direction, with $l = 0°$ in the direction of the Galactic Center. Longitude $l = 180°$ is the *anticenter* direction. Galactic longitudes 0° to 90° are called the *first quadrant*, I, as shown in Figure 8.6. This quadrant is part of the inner Galaxy, and is visible from the Northern Hemisphere. Longitudes $l = 90°$ to 180° correspond to the second quadrant (II), which is the outer Galaxy, also visible from the Northern Hemisphere. The third quadrant (III), $l = 180°$ to 270°, and fourth quadrant (IV), $l = 270°$ to 360°, are visible from the Southern Hemisphere. The Sun's path around the Galactic Center is called the *solar circle*. The Galaxy rotates in the clockwise direction in this coordinate system.

Objects farther from the Galactic Center than the Sun travel at a lower angular velocity around the center. This differential rotation means that our solar neighborhood is overtaking objects in the second quadrant. Although the Sun is always on the solar circle, and an object in quadrant II always has the same average distance from the Galactic Center, the object's distance to us decreases

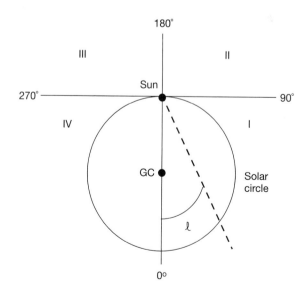

FIGURE 8.6 *Galactic longitude and galactic quadrants. The galaxy rotates clockwise. GC, Galactic Center.*

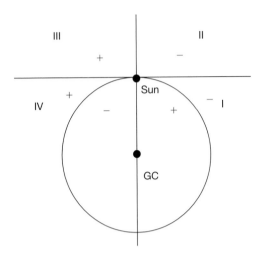

FIGURE 8.7 *Variation of Doppler shift in different Galactic quadrants. GC, Galactic Center.*

with time because of our different angular motions. Thus, there is a blueshifted Doppler line from objects in quadrant II, as indicated in Figure 8.7. Similarly, objects in quadrant III are slipping away from us with time, because their angular velocity is slower, so they show a redshift. Quadrants IV and I show blueshifts and redshifts, respectively, outside the solar circle.

Inside the solar circle, the pattern reverses: Objects move at a higher angular velocity than the Sun, so in the first quadrant they are moving away from us, and we measure a redshift. In the fourth quadrant they are moving toward us, so we measure a blueshift.

Beyond the solar circle, the velocities continuously increase or decrease with distance. For example, in the second quadrant the velocity becomes more and more negative (larger blueshifts) for more-distant objects, as shown in Figure 8.8. The radial velocity increases with distance in the third quadrant.

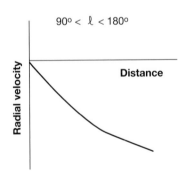

FIGURE 8.8 *Radial velocity (V_R) in the second Galactic quadrant.*

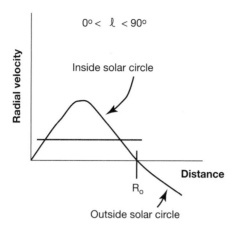

FIGURE 8.9 *Radial velocities in the first Galactic quadrant.*

In the first and fourth quadrants, the velocities are a bit more complicated because some points are closer to the Galactic Center than the Sun and some are
farther away. In Figure 8.9, the first quadrant shows positive radial velocities inside the solar circle and negative radial velocities beyond the solar circle. The
horizontal line indicates that there are two points inside the solar circle that
have the same radial velocity.

This duplicity results because a given line of sight intersects a circle twice, and
the component of velocity along the line of sight is identical at each intersection
because the two points are moving around the Galactic Center with the same angular velocity and direction cosines. In Figure 8.10, points a and b have different
space vectors, but they have the same radial velocity measured from the Sun.

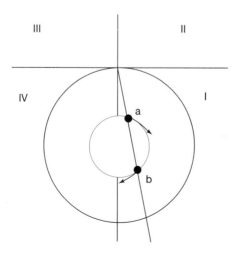

FIGURE 8.10 *Motions inside the solar circle.*

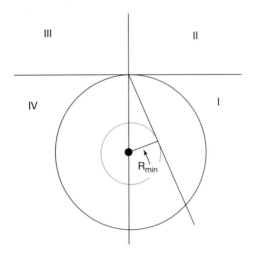

FIGURE 8.11 R_{min} occurs at a tangent point.

Note that there is one longitude that will intersect an inner circle at its tangent point (see Figure 8.11). At that longitude, there is no distance ambiguity for the observed velocity, because only the tangent point will have that particular radial velocity. At that point, the distance R_{min} from the Galactic Center is a minimum for that galactic longitude. The angle α used to define the space vector relative to our line of sight (as in Figure 8.5) is 0 since we are observing along a tangent, so the velocity (which is a maximum for that longitude) is given by

$$V_{R,max} = \theta(R_{min}) - \theta_o \sin(l)$$

For a flat rotation curve, this becomes

$$\boxed{V_{R,max} = \theta_o[1 - \sin(l)]}$$

between $l = 0°$ and $90°$.

8.7 Longitude-Velocity Diagrams

There is gas throughout the Galaxy, so we detect emission along every line of sight. Radio astronomers customarily plot the observed brightness of gas as a function of longitude and velocity rather than longitude and latitude. The resulting *longitude-velocity (l,v) diagram* provides some kinematic information as well as the distribution of observed objects.

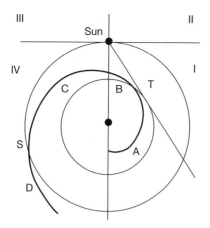

FIGURE 8.12   *Schematic spiral arm.*

Mapping spiral arms on (l,v) diagrams requires interpretation of Doppler-shifted motions in our Galaxy. As an example, consider the spiral in Figure 8.12, which is similar to the Sagittarius arm in quadrant I of our Galaxy; Figure 8.13 shows the corresponding (l,v) diagram for this spiral arm.

The (l,v) diagram was made from the equations given previously, assuming a distance of the Sun from the Galactic Center of 8.5 kpc and a circular velocity of 220 km s^{-1}. The right-hand loop from l = 0° to 40° (part A) is the arm inside the solar circle in the first quadrant. The high positive velocities are the region nearest the Galactic Center. The tangent point (T) is the point of maximum radial velocity, near l = 40° and v = 90 km s^{-1}. Part B is closer to the Sun, so the radial velocity approaches 0. The lower loop (C) is the arm in the fourth quadrant, with the arm crossing the solar circle at S, where velocity equals zero again. Outside the solar circle (D), the velocities become positive. This arm does not cross into quadrant II or III.

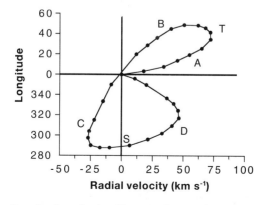

FIGURE 8.13   *Longitude-velocity diagram for quadrants I and IV.*

On (*l*,v) diagrams there is a *locus*, or line, of *terminal velocities* that arises because of tangent points. All of the emission near a tangent point will be at a similar velocity for a wide range of distances. The result is velocity *crowding*; that is, crowding of a lot of matter at different distances into the same velocity-longitude point. A plot of radial velocity versus longitude for terminal velocities is shown in Figure 8.14 for a circular velocity $\theta_o = 220$ km s^{-1} all the way to the center of the galaxy, which is unrealistic.

Figure 8.15 shows the CO emission in quadrant III of our Galaxy. Note the locus of terminal velocities running diagonally from $l = 340°$ to $300°$ for $V_{LSR} \sim -110$ to -40 km s^{-1} (marked A in the figure). The other side of the emission distribution, at low V_{LSR} in this diagram (B), corresponds to the solar neighborhood; there is no bright (nearby) emission from positive velocities in quadrant IV.

Some clouds with unusually high velocities of ~100 km s^{-1} or more and with the wrong sign for their galactic longitude (for example, blueshifted in quadrant I beyond the solar circle) have been identified as *high-velocity clouds* (HVCs). They cover more than 10% of the sky. Some of these HVCs may be part of a *galactic foun-*

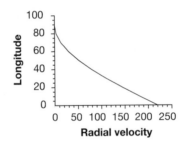

FIGURE 8.14   *Locus of terminal velocities in quadrant I.*

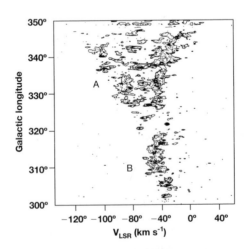

FIGURE 8.15   *^{13}CO (l,v) diagram for the Milky Way in quadrant III. (From Bronfman et al. 1988.)*

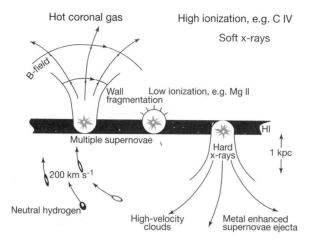

FIGURE 8.16   *This schematic diagram shows a galactic fountain, where material expands out of the plane and falls back down. (From Norman and Ikeuchi 1988.)*

tain, in which superbubbles of ionized gas extend out of the plane of the Galaxy. Supernovae create hot shocked gas that expands out of the plane and falls back down to the disk when it cools, as illustrated in Figure 8.16. Other HVCs result from a tidal interaction between the Milky Way and the Large and Small Magellanic Clouds, called the *Magellanic Stream*. HVCs are also observed in other galaxies.

8.8 Spiral Structure in the Milky Way

The spiral structure of the Milky Way is difficult to discern because of our internal vantage point. Use of different spiral tracers leads to different conclusions about the pitch angle and length of individual arms and spurs; overall, the Milky Way has an arm separation of 1–2 kpc and a pitch angle of 12°–25°.

There are several common methods for determining the distributions of stars and gas. The most direct way is to plot stars with distances determined from spectrophotometry, but only a relatively small fraction of stars have known distances. These distances typically are accurate to about 10%. In 1952, W. Morgan, S. Sharpless, and D. Osterbrock noted that OB associations, which are clusters of dozens of OB stars, outlined spiral arms in the Milky Way. Because O stars sometimes form in interarm regions, however, there is confusion in interpreting the main spiral patterns.

In the solar neighborhood, the star density is about 1 solar-type star per cubic parsec, dropping to ~0.2 pc^{-3} at a distance of 600 pc in the direction of the Galactic Center and rising again to 1 pc^{-3} at 2–3 kpc in the direction of the Galactic Center. The drop is probably the interarm region between us and the next inner spiral arm, which is the *Sagittarius-Carina arm*, and the rise at 2–3 kpc is the arm itself. The Sun is in the local *Orion spur*, which is not a well-developed spiral arm but a piece that has a higher pitch angle than the main arms; the *Perseus arm* is the nearest arm in the outer Galaxy.

Studies of open clusters with main sequence stars earlier than B3 show similar distributions. Cepheids with long periods >15 days also are concentrated in arms. Type A stars, early M giants, Be stars, Wolf-Rayet stars (which are high-mass stars in a helium-burning phase), carbon stars, and Type II pulsars also show slight concentrations to spiral arms. Most of these objects can be observed only locally because they are intrinsically faint. Figure 8.17 shows the local distribution of young spiral arm tracers.

Radio and optical H II regions generally follow the spiral pattern, as shown by Y.M. Georgelin and Y.P. Georgelin, although H II regions can be difficult to connect into spiral arms.

Supergiants follow the spiral arms but by themselves do not show spiral structure clearly. Early M stars (probably giants) are concentrated in the Norma-Scutum arm, whereas later M stars are not.

Kinematic distance techniques are applied to gaseous emission regions of H II, H I, and CO. The distance ambiguity can sometimes be resolved if there are additional associated tracers such as H II regions with spectrophotometric distances; errors in distances determined this way may be as high as 10% to 20%. The local distribution of CO is shown in Figure 8.18, and the distributions of giant atomic clouds and giant molecular clouds are shown in Figure 8.19.

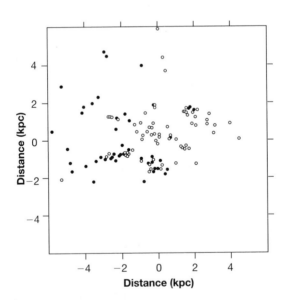

FIGURE 8.17   *Local distribution of long-period Cepheids. The Sun is located at (0,0). (From Becker and Fenkart 1971 [open circles] and Vogt and Moffat 1975 [closed circles]; figure adapted from Elmegreen 1985.)*

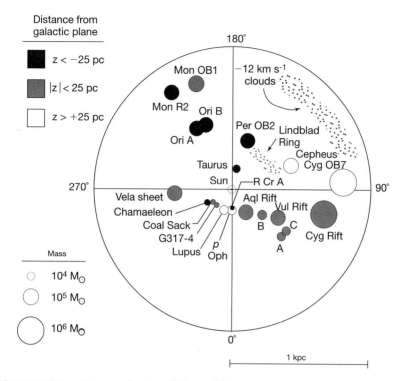

FIGURE 8.18 *CO near the Sun. (Adapted from Dame 1988.)*

The global fits to our Galaxy include either two or four main spiral arms, with pitch angles between 5° and 27°. For comparisons of models, it is useful to transform the spiral maps into (log r, θ) diagrams, as discussed in Chapter 5. Figure 8.20 shows a polar plot for the spiral tracers in our Galaxy. The H I in the outer Galaxy has two main spiral arms between log (radius, pc) = 4.1 and 4.2, whereas there are probably four main arms in the inner Galaxy between log (radius, pc) = 3.85 to 4.0.

Recent near-infrared studies by T. Matsumoto show a peanut-shaped bulge in our Galaxy, where there is also a velocity anomaly observed in 21 cm. L. Blitz and D. Spergel account for these features by a model with a small triaxial bar whose long axis is directed toward the Sun in the first quadrant, as shown in Figure 8.21.

The classification of the Milky Way is currently best described as SAB(rs)bc II, Arm Class M (multiple arm; see Chapter 5). The spiral galaxy NGC 1232, shown in Figure 8.22, is one of the galaxies most closely resembling the inferred shape of the Milky Way. There are multiple spiral arms, and there is a small bar in the center. NGC 1232A may be a companion dwarf galaxy (its radial velocity is uncertain), similar to our Magellanic Clouds.

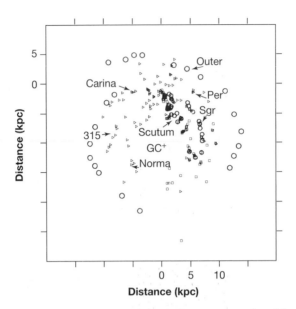

FIGURE 8.19   *The spiral structure of the Milky Way is outlined by several tracers. Open circles represent giant H I clouds, squares are giant molecular clouds, triangles are optical H II regions, and small boxes are radio H II regions. (Figure adapted from Elmegreen 1985.)*

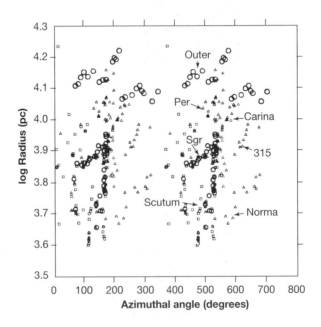

FIGURE 8.20   *A (log r, θ) plot of the Milky Way. (Elmegreen 1985.)*

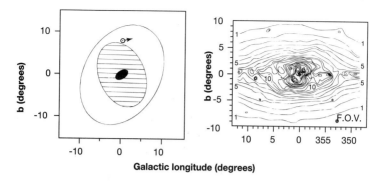

FIGURE 8.21 *(right) A 2.4μ image of the Galactic Center (Matsumoto et al. 1992), (left) modeled by Blitz and Spergel (1991); the black oval is the bar.*

FIGURE 8.22 *B band image of NGC 1232 and NGC 1232A. (Imaged with the Burrell-Schmidt telescope at KPNO by the author.)*

8.9 The Plane of the Milky Way

The plane of the Milky Way has been mapped in many wavelength bands, from high-energy gamma-ray to low-energy infrared. References are given at the end of the chapter for many groups involved in the surveys. Gamma-ray observations include all photons more energetic than 100 MeV, corresponding to frequencies of about 10^{14} GHz. There are compact gamma-ray sources in the Galactic Center that are coincident with pulsars; the bulk of the emission is uniformly spread along the plane and arises from collisions between cosmic rays and hydrogen clouds. In x-ray, the photons range from 0.25 to 1.25 keV, corresponding to frequencies of about 10^8 GHz. Some x-rays are associated with supernova remnants and hot, shocked gas. Figure 8.23 shows part of the galactic plane at optical, near-infrared, and infrared wavelengths.

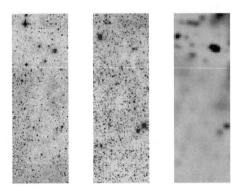

FIGURE 8.23 *A slice of the galactic plane in R (left), K (middle), and 12μ (right; IRAS). (Images by University of Massachusetts 2MASS team and Do Kester.)*

FIGURE 8.24 *The plane of the Milky Way at 1.25μ. (Imaged by COBE DIRBE; see Website listing for NIR.)*

The Digital Sky Survey (Space Telescope Science Institute) has mapped out the optical light of the plane in R band, corresponding to frequencies of about 8×10^5 GHz. Because of dust obscuration, most of these photons originate from stars within a kiloparsec or so of the Sun. Hα nebulosity also shows up in this passband and extends throughout the disk. Near-infrared emission at 1.2, 2.2, and 3.5μ, corresponding to frequencies of ~5×10^5 GHz, is dominated by light from K stars, as seen in Figure 8.24. The bulge is particularly prominent at these wavelengths. The infrared emission at 12, 60, and 100μ corresponds to frequencies of 3000–25000 GHz. The dominant source of this emission is reradiation from dust grains heated by starlight. Because dust and gas are well mixed in the interstellar medium, the distribution of the infrared emission strongly resembles the gamma-ray sky.

The atomic and molecular hydrogen distribution in the plane also resemble the infrared and gamma-ray emission distributions. The molecular hydrogen is traced by the ^{12}CO emission, at 143 GHz. In detail, it is clumped primarily into giant molecular clouds. There is also a strong concentration in the vicinity of the Galactic bulge, corresponding to the 4-kpc "ring." There are several filaments

FIGURE 8.25   *Grayscale intensities of 21-cm emission greater than 10° from the plane, with white lines representing polarization. (From Heiles and Jenkins 1976.)*

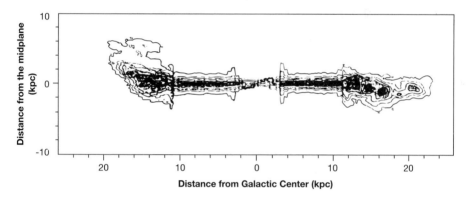

FIGURE 8.26   *H I distribution along the plane of the Milky Way, based on observations at galactic longitudes 75° (left side) and 255° (right side). (Figure from Burton 1992.)*

above the Galactic plane. The atomic hydrogen, at about 2 GHz, also shows large-scale clumping around the giant molecular clouds. There is a concentration in the bulge, and there is much clumpiness along the plane. The atomic gas shows much filamentary structure far from the plane, with loops and arcs that parallel magnetic field polarization maps, as shown in Figure 8.25.

The H I plane of our Galaxy is warped, with an increasing scale height with increasing galactocentric distance, as shown in Figure 8.26. Warps are common in other galaxies; an example is shown in Figure 8.27. Simulations show that warps occur in undisturbed galaxies from oscillations in the disks or in interacting galaxies from tidal effects.

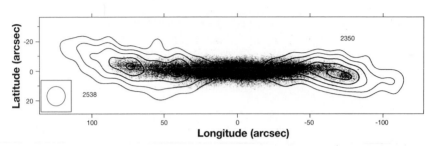

FIGURE 8.27 *H I velocity channel contours overlaid on a B band image of UGC 7170, showing the gas warp. (Figure from Cox et al. 1996.)*

Exercises

1. Determine the total peculiar velocity of a star with peculiar velocity components
   a. u = 21 km s^{-1}, v = 14 km s^{-1}, and w = 8 km s^{-1}
   b. u = 5 km s^{-1}, v = 3 km s^{-1}, and w = 80 km s^{-1}.
   What is the probable type of each star?
2. Suppose an H II region at a galactic longitude of 100° is observed to have a Doppler-shifted [OII] line at λ = 3727.625 Å (assume a rest wavelength of 3727.7 Å). What is the heliocentric distance of the H II region (determined from the radial velocity of the object)? What is the distance of the H II region from the Galactic Center?
3. From the (log r, θ) plot of the Milky Way in Figure 8.20, determine the pitch angles of several spiral arms in our Galaxy. Show the range of possibilities based on selecting different spiral features.
4. The midplane number density of stars is 2.51 × 10^{-8} pc^{-3} for O stars, 2.51 × 10^{-3} pc^{-3} for type F, and 6.3 × 10^{-2} pc^{-3} for type M. Determine the densities of each of these stars at z = 200 pc.
5. If a star at *l* = 210° is 4 kpc from the Sun, what is its distance from the Galactic Center?
6. Sketch an (*l*,v) diagram for an arm extending through quadrants II and III.
7. Complete the steps between the equations on page 171 deriving the radial velocity expression.

Unsolved Problems

1. What is the detailed structure of the Milky Way?
2. What is the history of star formation in the Milky Way?

Useful Websites

The Astrophysics Data Facility of the Goddard Space Flight Center, NASA, has produced a color poster of the Milky Way in several wavelengths. A poster and Web exploration of the data are available through http://adc.gsfc.nasa.gov/mw/milkyway.html

Details about the satellites and institutions are available through the following sites:
x-ray data via http://www.rosat.mpe-garching.mpg.de/survey/sxrb/
gamma-ray data via http://lhea.www.gsfc.nasa.gov/doc/gamcosray/EGRET/
egret.html and http://cossc.gsfc.nasa.gov/cossc/EGRET/html
optical data via http://www-gsss.stsci.edu/
NIR (1.25, 2.2, 3.5μ) data via http://ssdoo.gsfc.nasa.gov/astro/cobe/
cobe_home.html
CO data via http://adc.gsfc.nasa.gov/pub/adc/archives/catalogs/8/8039/
IR (12, 60, 100μ) data via http://www.ipac.caltech.edu/ipac/iras/iras.html and
http://ssdoo.gsfc.nasa.gov/astro/iras/iras_home.html

Further Reading

Albers, H. 1972. Infrared surveys of the southern Milky Way. 2. Suspected supergiants. *Astrophysical Journal* 176:623.

Baade, W., and N. Mayall. 1951. Problems of cosmical aerodynamics, Dayton: Central Air Documents Office. (See also Mihalas and Binney 1981.)

Bahcall, J. 1986. Star Counts and Galactic Structure. *Annual Review of Astronomy and Astrophysics*, 24:577.

Bahcall, J., and R. Soneira. 1980. The universe at faint magnitudes. I. Models for the galaxy and the predicted star counts. *Astrophysical Journal (Suppl.)* 44:73.

Becker, W., and R. Fenkart. 1971. A catalogue of galactic star clusters observed in three colours. *Astronomy and Astrophysics (Suppl.)* 4:241.

Binney, J. 1995. The evolution of our galaxy. *Sky and Telescope* 89:20.

Blitz, L., J. Binney, K. Lo, J. Bally, and P. Ho. 1993. The centre of the Milky Way. *Nature* 361:417B.

Blitz, L., and D. Spergel. 1991. Direct evidence for a bar at the galactic center. *Astrophysical Journal* 379:61.

Blitz, L., and D. Spergel. 1991. The shape of the galaxy. *Astrophysical Journal* 370:205.

Brand, J., and L. Blitz. 1993. The velocity field of the outer galaxy. *Astronomy and Astrophysics* 275:67.

Bronfman, L., M. Bitran, and P. Thaddeus. 1988. ^{13}CO in the southern galactic plane. In *Molecular clouds in the Milky Way and external galaxies*, ed. R. Dickman, R. Snell, and J. Young. 318. New York:Springer-Verlag.

Burstein, D., and C. Heiles. 1982. Reddenings derived from H I and galaxy counts: Accuracy and maps. *Astronomical Journal* 87:1165.

Burton, W. B. 1985. Leiden-Green Bank survey of atomic hydrogen in the galactic disk. I. l,v and b,v maps. *Astronomy and Astrophysics (Suppl.)* 62:365.

Burton, W.B. 1992. Distribution and observational properties of the ISM. In *The Galactic Interstellar Medium: Saas-Fee advanced course 21*, ed. D. Pfenniger and P. Bartholdi, 1. New York: Springer-Verlag.

Caldwell, J., and J. Ostriker. 1981. The mass distribution within our galaxy — A three-component model. *Astrophysical Journal* 251:61.

Cox, A., L. Sparke, G. van Moorsel, and M. Shaw. 1996. Optical and 21-cm observations of the warped, edge-on galaxy UGC 7170. *Astronomical Journal* 111:1505.

Dame, T. 1988. A composite CO survey of the entire Milky Way. In *Molecular clouds in the Milky Way and external galaxies*, ed. R. Dickman, R. Snell, and J. Young. New York: Springer-Verlag. 309.

Dame, T. 1988. The molecular Milky Way. *Sky and Telescope* 76:22.

de Vaucouleurs, G., and W. Pence. 1978. An outsider's view of the galaxy—Photometric parameters, scale lengths, and absolute magnitudes of the spheroidal and disk components of our galaxy. *Astronomical Journal* 83:1163.

Elmegreen, D. M. 1985 Spiral structure of the Milky Way and external galaxies. In *IAU Symposium 106: The Milky Way Galaxy*, ed. H. van Woerden, R. J. Allen, and W. B. Burton. 255. Dordrecht: Reidel.

Freeman, K. 1987. The galactic spheroid and old disk. *Annual Review of Astronomy and Astrophysics* 25:603.

Fresneau, A., A. Acker, G. Jasniewicz, and M. Piat. 1996. Kinematical search in the optical for low-mass stars of the Gould Belt system. *Astronomical Journal* 112:1614.

Georgelin, Y. M., and Y. P. Georgelin. 1976. The spiral structure of our galaxy determined from H II regions. *Astronomy and Astrophysics* 49:57.

Gilmore, G., I. King, and P. van der Kruit. 1989. *The Milky Way as a Galaxy: Nineteenth Saas-Fee Course*, ed. R. Buser and I. King. Geneva: Geneva Observatory.

Gilmore, G., R. Wyse, and K. Kuijken. 1989. Kinematics, chemistry, and structure of the Milky Way. *Annual Review of Astronomy and Astrophysics* 27:555.

Hauser, M., T. Kelsall, D. Leisawitz, and J. Weiland. 1995. COBE DIRBE. *NASA publication 95-A, Goddard Space Flight Center*, Greenbelt, MD: National Aeronautics and Space Administration Publication.

Heiles, C., and E. Jenkins. 1976. An almost complete survey of 21-cm line radiation for galactic latitudes of 10 degrees and higher. V. Photographic presentation and qualitative comparison with other data. *Astronomy and Astrophysics* 46:333.

Kahn, F. 1994. Galactic fountains. *Astrophysics and Space Science* 216:325.

Kerr, F., P. Bowers, M. Kerr, and P. Jackson. 1986. Fully sampled neutral hydrogen survey of the southern Milky Way. *Astronomy and Astrophysics (Suppl. Ser.)* 66:373.

Mathewson, D., and V. Ford. 1971. Polarization observations of 1800 stars. *Monthly Notices of the Royal Astronomical Society* 153:525.

Matsumoto, T., S. Hayakawa, H. Koizumi, H. Muakami, K. Uyama, T. Yamagani, and J. Thomas. 1992. Balloon observation of the central bulge of our galaxy in near infrared radiation. In *The Galactic Center*, ed. G. Riegler and R. Blandford, 48. New York: American Institute of Physics.

Mihalas, D., and J. Binney. 1981. *Galactic astronomy: Structure and kinematics*, San Francisco: W.H. Freeman.

Norman, C., and S. Ikeuchi. 1988. The disk-halo connection and the nature of the interstellar medium. In *Kerr Symposium of the Outer Galaxy*, ed. L. Blitz and F. Lockman, 115. Berlin: Springer-Verlag.

Pöppel, W. 1997. The Gould Belt system and the local interstellar medium. *Fundamentals of Cosmic Physics* 18:1.

Scheffler, H., and H. Elsässer. 1987. *Physics of the galaxy and interstellar matter.* New York: Springer-Verlag.

Schulman, E., J. Bregman, and M. Roberts. 1994. An H I survey of high velocity clouds in nearby disk galaxies. *Astrophysical Journal* 423:180.

Snowden, S., et al. 1995. First maps of the soft X-ray diffuse background from the ROSAT XRT/PSPC All-Sky Survey. *Astrophysical Journal* 454:643.

Tammann, G. 1970. The Milky Way in comparison to other galaxies. In *IAU Symposium 38, The Milky Way Galaxy*, ed. W. Becker and G. Contopoulos, 236. Dordrecht: Reidel.

Unavane, M., R. Wyse, and G. Gilmore. 1996. The merging history of the Milky Way. *Monthly Notices of the Royal Astronomical Society* 278:727.

Vallee, J. 1995. The Milky Way's spiral arms traced by magnetic fields, dust, gas, and stars. *Astrophysical Journal* 454:119.

van Woerden, H., R.J. Allen, and W.B. Burton eds. *IAU Symposium 106: The Milky Way Galaxy*, Dordrecht: Reidel.

Vogt, N., and A. Moffat. 1975. Galactic structure based on young southern open star clusters. *Astronomy and Astrophysics* 20:85.

Wheelock, S., et al. 1994. *IRAS Sky Survey atlas*. JPL Publication. 94–11, Pasadena: Jet Propulsion Laboratory.

CHAPTER 9

Spiral Wave Kinematics

Chapter Objectives: to understand small disk perturbations and their effect on stars and gas

---

Toolbox:

epicyclic frequency	pattern speed
density waves	dispersion relation
resonances	bar perturbation
modes	WASER
corotation	stochastic star formation

9.1 Epicycles

In spiral galaxies, stars and gas in the disk travel in nearly circular orbits (unless they are near a strong bar, as described in Section 9.5). They also wobble in their orbits because of the perturbations due to the combined gravitational effects of other objects and also perhaps due to initial, random velocity components of the molecular cloud from which they formed. These wobbles to the circular motion are periodic and, in a comoving reference frame, trace out ellipses called *epicycles* as the stars move inward and outward in the disk. The frequency of oscillation is known as the *epicyclic frequency*.

The wobbling motions in a disk are dictated by the conservation of angular momentum, which operates through the Coriolis force. If a star is going slower than the average (i.e., Local Standard of Rest, LSR) motion at that radius, it falls inward because its centrifugal force is too weak to balance the gravitational force of the inner galaxy. It does not just fall radially inward; because the system is rotating, it accelerates to a faster angular speed as it moves inward. This Coriolis acceleration is $-2\omega \times v$ for angular velocity ω of the rotating reference system and velocity v of the object; recall that the $\times$ represents a cross-product, and the acceleration is perpendicular to both the angular velocity and the instantaneous space velocity. When the star falls inward, the Coriolis acceleration is directed forward in the orbit, so the star increases its circular (i.e., tangential) speed. This motion may also be viewed as a result of the conservation of angular momentum $L = mv_\theta r$ for mass m, radius r, and tangential velocity v_θ; a decreasing radius leads to an increasing tangential velocity. The increasing tangential velocity causes the centrifugal force to exceed the inward gravitational force, plus it leads to an outward Coriolis force, so the star moves outward again. As a result of the balance between Coriolis, centrifugal, and gravitational forces, the star oscillates in a regular fashion inward and outward, forward and backward in its orbit.

Viewing from outside the galaxy (that is, from an inertial, or fixed, reference frame), we would see oscillations of the star's orbit, while the LSR would move in a circle, as shown in Figure 9.1 on the left. In the reference system that is co-moving with the LSR, we would see the star trace out an ellipse, as in Figure 9.1 on the right.

An expanded view of the ellipse is shown in Figure 9.2. Note that the epicycle's rotation is retrograde; that is, in the direction opposite the rotation of the galaxy. The coordinates for the small-scale perturbed motions will be represented by ξ and η for the radial and azimuthal components, respectively.

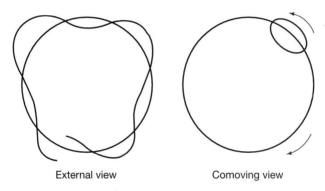

External view   Comoving view

FIGURE 9.1   *Epicycles in external and comoving reference frames. The arrows represent the relative directions of rotation of the disk and epicycle.*

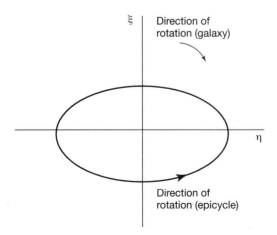

FIGURE 9.2   *Epicycle coordinate system. Note that the direction of motion in the epicycle is retrograde with respect to the galaxy rotation.*

In a reference frame rotating at the LSR angular velocity ω_0, the radial acceleration is given by the time derivative of the radial velocity v_r according to Newton's law:

$$m\frac{dv_r}{dt} = F_r + m\omega_0^2 r + 2m\omega_0 v_\theta$$

where v_θ is the tangential velocity component. The first term, F_r, is the gravitational force that determines galaxy rotation:

$$F_r = -m\omega_{gal}(r)^2 r$$

where $\omega_{gal}(r)$ is given by the rotation curve. The second term is from centrifugal force, and the third term is from the Coriolis force in the rotating reference system. The mass times the acceleration in the tangential direction is given by

$$m\frac{dv_\theta}{dt} = F_\theta - 2m\omega_0 v_r$$

where the first term represents the force in the tangential direction, which is 0 for an object in a symmetric galaxy, and the second term is the Coriolis force in the tangential direction.

Now consider the effect of small perturbations to the uniform circular motion, which lead to the epicyclic motions. A small perturbation in radius changes the position from some starting value r_0 to $r_0 + \xi$. The gravitational force varies as the star moves inward and outward, so the angular velocity of the LSR varies too. The angular velocity of the LSR can be expanded by a Taylor's series to represent its variation. Its expansion, keeping first-order terms, is

$$\omega_{gal} = \omega_0 + \frac{\partial \omega_{gal}}{\partial r}(r - r_0) = \omega_0 + \frac{\partial \omega_{gal}}{\partial r}\xi$$

We can plug this expression back into the radial acceleration equation. In Chapter 8, we described Oort's A, which is given by $A = -\frac{1}{2}r\,(d\omega_{gal}/dr)$. We will use it here to rewrite the equation (after some algebra) as

$$m\frac{dv_r}{dt} = 4mA\omega_0\xi + 2m\omega_0 v_\theta$$

We also have the tangential term from before,

$$m\frac{dv_\theta}{dt} = -2m\omega_0 v_r$$

These equations are linear in the velocities and positions, so the solutions are sine and cosine functions:

$$v_r = v_{r_0}\sin(\kappa t)$$
$$v_\theta = v_{\theta_0}\cos(\kappa t)$$
$$\xi = -[v_{r_0}/\kappa]\cos(\kappa t) = -\xi_0\cos(\kappa t)$$
$$\eta = [v_{\theta_0}/\kappa]\sin(\kappa t) = \eta_0\sin(\kappa t)$$

The value of κ, the epicyclic frequency, is determined by solving the equations simultaneously. The result is:

$$\kappa^2 = 4\omega_{gal}^2\left[1 + \frac{1}{2}\left(\frac{r}{\omega_{gal}}\frac{d\omega_{gal}}{dr}\right)\right]$$

As an example of typical values, consider our own Galaxy. Assume 8.5 kpc for the distance of the Sun from the Galactic Center and 220 km s^{-1} for the rotation velocity at that distance. These give an epicyclic period p = $2\pi/\kappa \sim 2 \times 10^8$ yr near the Sun, for an epicyclic frequency of 36 km s^{-1} kpc^{-1}. Note these unusual units, which are used because they relate directly to observed motions in the disk, commonly measured in kilometers per second, and distances from the center of a galaxy, commonly expressed in kiloparsec. Of course, the frequency could also be expressed in inverse seconds or inverse years.

The distance a star moves during its epicycle is given by the ratio of the peak velocity within the epicycle (which is about the same as the peculiar velocity) to the frequency: $v_{peculiar}/\kappa$. For $v_{peculiar} \sim 30$ km s^{-1}, this distance corresponds to about 1 kpc. The ratio of the azimuthal and radial axes of the epicycle is given by ratio of the amplitudes of these perturbations, which is

$$\frac{\eta_0}{\xi_0} = \frac{2\omega_0}{\kappa}$$

In the Milky Way at the radial position of the Sun, this ratio is ~1.5, because the angular velocity is 26 km s^{-1} kpc^{-1} and the epicyclic frequency is 36 km s^{-1} kpc^{-1}. This ratio can also be expressed in terms of Oort's constants A and B:

$$2\omega_0/\kappa = \sqrt{-B/(A - B)}$$

The epicyclic frequency similarly can be written as

$$\kappa = \sqrt{-4B(A - B)}$$

For every complete revolution around the Galaxy, the Sun completes about 1.4 epicycles (given by the ratio of κ/ω_0).

Stars also have motions perpendicular to the disk, so we write similar equations for the vertical direction displacement:

$$m\frac{d^2z}{dt^2} = F_z$$

In this case, the frequency is the *vertical frequency* ν rather than an epicyclic frequency κ, and it depends on the vertical gravitational potential: $\nu \approx \sqrt{4\pi G \rho_0}$ for average disk density ρ_0. In the solar neighborhood, this frequency is 3.2×10^{-15} s^{-1}. The vertical oscillation period is therefore $2\pi/\nu \approx 6.5 \times 10^7$ yr. A consequence of these two being noncommensurate and also not integral multiples of the orbital frequency is that the orbits are open (like rosettes) rather than closed orbits as in our Solar System.

Details of stellar dynamics are beyond the scope of this book. They may be found in references such as Binney and Tremaine 1987.

9.2 Spiral Density Waves

The process of generation and maintenance of spiral structure was a mystery 50 years ago. Galaxies have rotated 20 to 100 times in the lifetime of the Universe, which means that if the same stars and gas always make up the spiral structure, spiral arms should be wound up 20 to 100 times. Such arms are called *material arms* and the problem of wrapping is known as the *winding dilemma*. Real galaxies never have arms wrapped in azimuth more than once or twice. This observation leads to the conclusion that either spiral arms are short-lived or they are not material arms. In Chapter 5 we saw that grand design galaxies have spiral patterns in old as well as in young stars; this observation suggests that their arms are not purely young.

A partial solution to the winding dilemma is given by the *spiral density wave* theory. The mathematical formulation for the density wave model was developed by C. C. Lin and Frank Shu in the 1960s, following some initial ideas by Bertil Lindblad in the 1940s. A spiral density wave is a gravitational perturbation that propagates through a disk. The effect of the wave is to pile up material temporarily at the wave crest, and this piling up makes the spiral arms. An analogy is a hill in the road. Cars will bunch up temporarily as they climb the hill, then move on, while more cars bunch up behind them. Similarly, stars and gas bunch up in the wave crests. This density concentration causes a deflection of the nearby stars and gas as they approach and move through the wave. Arms caused by density waves are not material arms, and they do not wrap up as quickly as material arms do.

The direction of rotation with respect to the winding of the arms is such that they would wrap up more and more with time, as if they were coiled springs being wound up tighter and tighter. Galaxies therefore have *trailing arms*. This is the sense of the winding in almost all disks. In Figure 4.3, for example, we infer that M51 is rotating counterclockwise. We see approximately equal numbers of galaxies with clockwise and counterclockwise wrapping, presumably as a result of their random orientations in the sky. In rare cases, possibly having to do with interactions, some arms have an orientation with respect to the galaxy rotation that has the sense of "unwinding"; these arms are said to be *leading*.

Epicyclic motions exist even without a spiral density wave (SDW), as discussed above. If we are in a reference frame rotating at a rate of $\omega - \kappa/2$, the star orbit appears to be a closed ellipse. If a density wave is present, then the gravity from the wave can deflect the stars and gas so their orbits will be closed ellipses in a rotating frame for a wide range of radii. This rotating frame is the "corotation" frame for the spiral wave; see Figure 9.3.

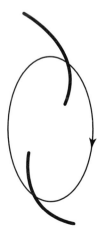

FIGURE 9.3 *Moving at the pattern speed, we would see stars move in closed elliptical orbits.*

The orbital ellipses attempt to precess with respect to the wave pattern, and gravity from the spiral density wave attempts to hold them in check; this action is called *gravity forcing*. A resulting possible configuration for the ellipses is shown in Figure 9.4. There is *orbit crowding* near the major axis of each ellipse, which means that stars bunch up at these points and create the spiral arms.

Consider what happens as a star approaches a density wave crest: Inside corotation, it is moving faster than the density wave pattern, so it approaches the arm from the trailing (i.e., concave) side and moves through. As the star approaches the arm, gravity draws it in because the arm has a higher density than the rest of the disk. The result is that the star is pulled to a larger radius, so its azimuthal speed decreases through the conservation of angular momentum. There is no pull in the middle of the arm. As the star moves through the arm, it is again pulled by the arm, so it goes to a smaller radius and it speeds up in azimuth. This effect is called *velocity streaming*. As a result, stellar velocities tend to be parallel to the arm inside an arm and at a high pitch angle in the interarm region. A sketch of this effect is shown in Figure 9.5.

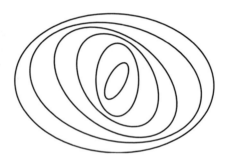

FIGURE 9.4   *Gravity forcing from an SDW causes the stellar ellipses to take on a regular pattern. Orbit crowding creates spiral arms.*

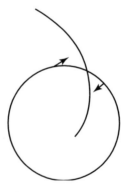

FIGURE 9.5   *A star is pulled from its orbit by a spiral arm.*

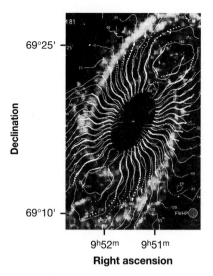

69°25'

Declination

69°10'

9ʰ52ᵐ   9ʰ51ᵐ

Right ascension

FIGURE 9.6 *Spider diagram for M81 shows wiggles due to velocity streaming along spiral arms. The lines show constant velocity; across a spiral arm, objects are deflected so they slow down and speed up. This effect causes the rotation to be noncircular, so there are kinks in the isovelocity lines. Such motion was predicted for this galaxy by Gottesman and Weliachew (1975). Thick white lines are from theory; thin white lines are 21-cm observations. (From Visser 1980.)*

The grand design spiral galaxy M81 provided the first evidence for a spiral density wave in other galaxies, on the basis of radio data showing velocity streaming along the arms. On a spider diagram (described in Chapter 6), velocity streaming shows up as C-shaped wiggles in the isovelocity contours because of the slowing down and speeding up of objects as they approach and leave an arm. This streaming can be seen in the H I gas in Figure 9.6.

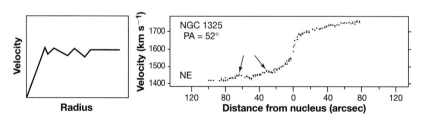

FIGURE 9.7 *(left) Streaming motions along an arm show up as local peaks and troughs on this schematic rotation curve. (right) Streaming motions in the rotation curve of the spiral galaxy NGC 1325, indicated by arrows. (From Rubin et al. 1982.)*

Velocity streaming has since been observed in many grand design galaxies. This streaming effect is also observed in star motions in the Milky Way. Streaming shows up as wiggles on a rotation curve, as mentioned in Chapter 7. The velocity is typically 10–20 km s^{-1} lower on the inner edges than on the outer edges of spiral arms; observations and a sketch of this effect are shown in Figure 9.7.

9.3 Resonances

If a star has an angular velocity ω and the pattern has an angular velocity ω_p, then the relative speed of the star with respect to the pattern is given by the difference, $\omega - \omega_p$. When the epicyclic frequency is synchronous with the relative motion of the spiral pattern in the absence of gravitational forcing by the wave, a condition known as a *resonance* is created. That is, a resonance occurs when the difference between the angular velocity and the pattern speed is an integral multiple of the epicyclic frequency:

$$\omega - \omega_p = \pm \frac{\kappa}{m}$$

for integer values of m, which is the number of arms. For example, a galaxy with m = 2 dominating would have two main arms.

In a two-arm galaxy, the two fundamental resonances are known as the *Inner* and *Outer Lindblad Resonances* (ILR, OLR), where, in the absence of forcing, one epicycle is completed in the time it takes to go from one arm to the next. That is, in the frame of reference rotating at ω_p, two epicycles are completed in one revolution around the galaxy. Here,

$$\omega - \omega_p = -\kappa/2 \text{ at OLR}$$
$$\omega - \omega_p = \kappa/2 \text{ at ILR}$$

At a resonance radius, the star is at the same point in its epicycle each time it encounters a spiral arm, so it receives the same gravitational pull at the same point in the epicycle. A familiar example of a resonance is when a swing is pumped at just the right rate to make it go higher and higher. In an analogous way, at a resonance radius in a galaxy, a star soaks up even infinitesimally weak wave energy and goes faster and faster in bigger and bigger epicycles. The ILR and OLR therefore define the ultimate limits to spiral arm structure, since a stellar density wave with finite amplitude cannot propagate beyond either limiting radius.

A more detailed look at the physics behind spiral density waves is given in Appendix 4, in which a dispersion relation for a density wave is derived from basic hydrodynamic equations. Solutions for the waves, leading to inward and outward motions in the disk, resonances, and subsequent density variations are considered.

In order to examine resonances in a disk, let us consider a simplified flat rotation curve, as shown in Figure 9.8. The angular velocity, $\omega = v/r$, and the epicyclic frequency, κ, for this rotation curve are plotted as functions of radius in Figure 9.9. The pattern's angular velocity is constant with radius, so it is represented as a straight horizontal line. Inside corotation, stars and gas move faster than the pattern, and outside corotation their angular velocity is slower. In Figure 9.9, the pattern speed is 20 km s^{-1} kpc^{-1}; this equals the angular velocity of the disk at a radius of about 7.5 kpc.

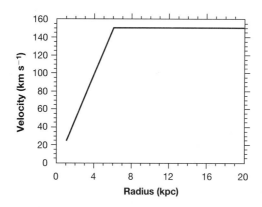

FIGURE 9.8 *A simplified rotation curve for a galaxy.*

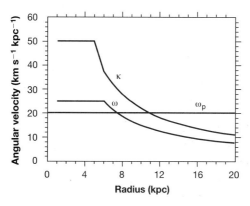

FIGURE 9.9 *Angular velocity curve, showing epicyclic frequency and v/r.*

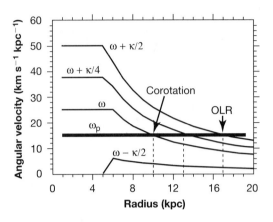

FIGURE 9.10   *Angular velocities for the rotation curve in Figure 9.8.*

We can plot curves for other differences between angular velocity and epicyclic frequency, as shown in Figure 9.10. These curves are useful for determining the locations of different resonances, which depend on the pattern speed. Let us consider where resonances would be located in this galaxy. Suppose it has a size $R_{25} = 17$ kpc. Furthermore, suppose that the arms end at the OLR, and that OLR is at R_{25}. We know that the pattern speed is equal to $\omega + (\kappa/2)$ at OLR. Thus, the intersection of the $\omega + (\kappa/2)$ line and the vertical line drawn at 17 kpc fixes the pattern speed, which in this case is 15 km s^{-1} kpc^{-1}, as indicated in the figure. Then the location of corotation (where $\omega_p = \omega$) is given by the intersection of the line for the pattern speed and the curve for ω, which here is at 10 kpc. The outer 4:1 resonance is where $\omega - \omega_p = \kappa/4$, so this radius is fixed by where the pattern speed intersects the 4:1 curve, at about 13 kpc in the figure. Note that this galaxy has no ILR, since the $\omega - (\kappa/2)$ line lies everywhere below the pattern speed line.

In the Milky Way, for the Sun at 8.5 kpc and $v_o = 220$ km s^{-1}, the ILR is estimated to be at ~3 kpc, with corotation at ~14 kpc and OLR at ~20 kpc. The pattern speed of the spiral density wave, ω_p, is estimated to be ~15 km s^{-1} kpc^{-1}.

The ratios of resonance radii vary for different rotation curves. For a power law fit to a rotation curve, where $v(R) \propto R^\alpha$, the ratio of a resonance radius R_{res} to the corotation radius R_{cor} is given by

$$\frac{R_{res}}{R_{cor}} = \left(1 - \frac{2}{m}\left[\frac{1+\alpha}{2}\right]^{1/2}\right)^{1/[1-\alpha]}$$

where $m = 2$ at ILR, -2 at OLR, and so on. If the galaxy has a flat rotation curve, then $\alpha = 0$, so $R_{res}/R_{cor} = (1 - (2/m)\sqrt{1/2})$; then $m = -2$ gives $R_{OLR}/R_{cor} = 1.7$. This approximation is probably not valid in the vicinity of the ILR because of deviations from the power law approximation to the rotation curve.

A more general representation of rotation curves is

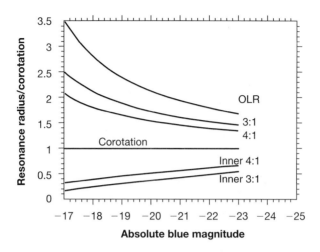

FIGURE 9.11   *Theoretical resonance ratios as a function of absolute magnitude.*

$$v(R) = \frac{R}{(R^A + R^{(1-\alpha)})}$$

for small A and α. This form has a nearly solid body rising part ($v \sim R^{(1-A)}$) at small R, and a nearly flat part ($v \sim R^{\alpha}$) at large R. Typically $\alpha \sim 0$ (which is a flat rotation curve) or a small value such as ± 0.1. The overall form is insensitive to A (which is usually taken as zero), which primarily affects the location of the ILR.

The ratios of resonances can be plotted as a function of absolute magnitude for the above rotation curve, as shown in Figure 9.11, because the slope α of the rotation curve scales with absolute magnitude to first approximation (see Section 7.3). Note the interesting result that resonances inside corotation are crowded together in bright galaxies but are more spread out in fainter galaxies.

9.4 Resonance Features

If a galaxy's rotation curve and any one resonance are known, then all of the other resonances as well as the pattern speed are precisely specified. There are several prominent kinds of features that might be the result of resonances. For example, we think that bright stellar spiral arms should end near the OLR, and bright star formation ridges in the arms end near corotation. If several such resonance features are seen in any one galaxy, and the ratios of their radial locations agree with theoretical resonance ratios, then the pattern speed can be determined.

FIGURE 9.12   *NGC 5248 has a bright inner ridge and fainter outer arms. (Imaged with the KPNO Burrell-Schmidt telescope by the author.)*

Consider several possible resonance features. Spiral arms extend to the apparent galaxy edge, generally around R_{25}. This arm end is assumed to be close to the OLR, because that is a consistent resonance radius according to rotation curve and pattern speed fits to several grand design galaxies. Most nonflocculent galaxies have two inner symmetric arms that extend to mid-disk, beyond which point bifurcations and spurs are more common. Some galaxies have prominent *ridges,* which are bright inner disks with apparently sharp edges (see, for example, the main inner region of M51 in the I band photo of Figure 4.3 and NGC 5248 in Figure 9.12).

These ridges and inner symmetric arms end near corotation, on the basis of observations of galaxies with known rotation curves and pattern speeds. In addition, there are often spurs, or short spiral arm pieces, that are symmetrically

FIGURE 9.13   *NGC 628 has short spurs branching off the main spiral arms. (Image from the STScI Digital Sky Survey.)*

FIGURE 9.14 *NGC 1433 has an inner ring and outer arms that form a ring. (Image from the STScI Digital Sky Survey.)*

placed between two main spiral arms, such as in NGC 628, shown in Figure 9.13, and NGC 5457, shown in Figure 9.18 (middle). In theory, these features occur at a resonance radius where four epicycles are completed in one complete orbit in the pattern reference frame (that is, where $\omega - \omega_p = +\kappa/4$, also called the ultra-harmonic or 4:1 resonance).

Rings are also very good resonance indicators. The observed ratios of outer to inner ring radii are nearly the same as the theoretical ratio of OLR to corotation and are reproduced in n-body simulations of ring formation. A galaxy with inner and outer rings is shown in Figure 9.14. Nuclear rings appear to form at the ILR, according to theory; an example is shown in Figure 9.15.

Dust lanes may also be useful resonance indicators. Dust lanes indicate the location of shocks in the interstellar medium as a result of supersonic gas motions relative to the arm. Further discussion of shocks is presented in Appen-

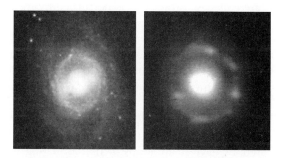

FIGURE 9.15 *(left) B band image of the barred spiral NGC 3351; note its inner ring. (From the Digital Sky Survey). (right) K band image of the center of NGC 3351 showing an ILR ring, which corresponds to the white solid region in the B band image. (Imaged with the KPNO 2.1-m telescope by D. Elmegreen and F. Chromey, Vassar College.)*

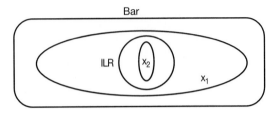

FIGURE 9.16 *Orientation of star orbits inside a bar, which is indicated by dots. Inside the ILR, orbits are perpendicular to the long axis of bar (designated by x_2); outside ILR, orbits are parallel (designated by x_1).*

dix 4. According to theory, dust lanes inside corotation are on the concave (inside) edge of spiral arms, whereas outside corotation they are on the convex (outside) edge of spiral arms; the crossover point, if there is one, should be near corotation.

Bars appear to end between the inner 4:1 resonance and corotation, according to computer models. In early-type barred spiral galaxies, the 4:1 resonance is very close to corotation, since these galaxies tend to be bright (refer to Figure 9.11). In weakly barred galaxies (which are later Hubble types, with fainter absolute magnitudes), the 4:1 resonance is farther from corotation. As we saw in Chapter 5, these exponential bars are smaller with respect to R_{25} than are flat bars.

Stellar orbits are strongly perturbed in the vicinity of a bar, as shown by G. Contopoulos, E. Athanassoula, F. Combes, and others. Computer simulations of how stars move in the presence of the strong noncircular gravitational potential indicate that orbits are highly elongated. Figure 9.16 shows the elongation of orbits in a bar, which is indicated by dots: Inside the ILR, orbits are perpendicular to the bar; these are called x_2 orbits. Outside the ILR, orbits are parallel to the bar; these are called x_1 orbits. Gas is perturbed by the bar too; the bar torque causes gas to fall in along the bar and collect at the ILR, as noted in Chapter 6. Inside the ILR, the bar torque has the opposite effect and pushes gas out to the ILR. Thus, rings of gas (and subsequent star formation) are often prominent in a barred galaxy and are a signature of an ILR. These models are based on a linear analysis although the bar perturbations are nonlinear, so the effects are merely illustrative.

In general, the orbits are not simple ellipses but are complicated figures. Some repeat their pattern periodically; others change with time. Orbits may be box-shaped, pointed, or have many loops in the corotating frame of the bar. Some of the common types of orbits are shown in Figure 9.17, based on n-body simulations.

Table 9.1 summarizes prominent features that may occur in a galaxy and their likely associated resonances.

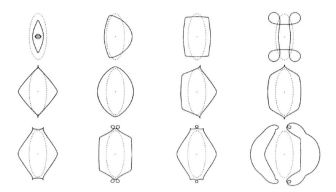

FIGURE 9.17 *Typical star orbits in a bar potential, based on n-body simulations (Athanassoula 1992.)*

TABLE 9.1 *Optical resonance indicators*

Feature	Resonance	Example (Figure number)
Outer two bright arms	OLR	M81 (5.12)
Inner ridge of star formation	Corotation	N5248 (9.12)
Dust lanes switching from concave side to convex side	Corotation	N1300 (5.9)
Spurs midway between two main arms	4:1	N628 (9.13)
Three-arm structure	Inner 3:1 to outer 3:1	M101 (9.18)
Nuclear ring	ILR	N3351 (9.15)
Inner ring	4:1	N1433 (9.14)
Bar	Between 4:1 and corotation	N1300 (5.9)

9.5 Symmetries

One method for assessing the twofold symmetry in a galaxy involves computer manipulations. First, an image is rotated 180° and subtracted from the original image. The difference is then truncated so that all negative pixels are set equal to zero, and the truncated difference is subtracted from the original image. The result contains only the parts of the original image that are twofold-symmetric. Many grand design galaxies have very prominent symmetric structures over most of their disks.

Three-arm symmetries can be measured by rotating images 120° and 240°, subtracting each from the original and truncating as before, and then subtracting both of the differences from twice the original. Multiple-arm galaxies sometimes have prominent three-arm structure over part of their disks, between the inner and outer Lindblad resonances for three-arm modes (i.e., at the two radii in the disk where three epicycles are completed once around the galaxy, $\omega - \omega_p = \pm\kappa/3$). In a three-arm structure, this means that one epicycle is completed in the time it takes to go from one arm to the next. The fact that the three-arm structure extends all the way between resonances implies that this wave is not a mode, because it gets absorbed at its own resonances. A possibility is that the two-arm mode excites a three-arm structure as well.

M101 is a classic multiple-arm galaxy. Figure 9.18 shows enhanced and symmetric versions of its image. A sketch of two-arm and three-arm extent is shown in Figure 9.19.

The outer circle locates the OLR. The three-arm structure extends between the inner and outer 3:1 resonances, and the two-arm structure extends between

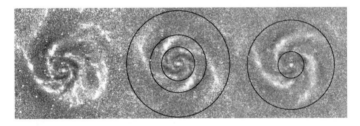

FIGURE 9.18 *M101 (NGC 5457) is shown in an enhanced image (left), with two-arm symmetric structure (middle), and three-arm symmetric structure (right). Circles in the middle image represent OLR, corotation, and inner 4:1; circles in the right image represent outer and inner 3:1 radii. (From Elmegreen, Elmegreen, and Montenegro 1992.)*

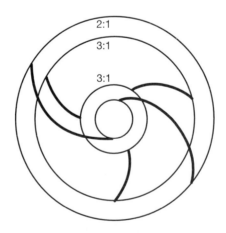

FIGURE 9.19 *Two-arm structure extends over a larger part of a disk than does three-arm structure because of resonances that limit the waves. 2:1, OLR; 3:1, inner and outer m = 3 resonances.*

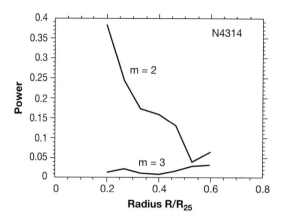

FIGURE 9.20 *Fourier components of NGC 4314 for K band data. The m = 2 component dominates in this two-arm galaxy.*

the inner and outer 2:1 resonances. Note that the three-arm extent is smaller than the two-arm extent.

Another method for determining symmetries in galaxies is to make use of *Fourier transforms*. The intensity at each radius in an image can be represented by a Fourier integral over azimuthal angle θ:

$$ I_m(r) = \int_{\theta=0}^{2\pi} I(r,\theta)e^{im\theta}d\theta $$

where m is the number of arms. A plot of the sum of the squares of the real and imaginary parts of $I_m(r)$ versus radius shows which modes dominate, as shown in Figure 9.20. In general, the m = 2 component is strongest. The radial extent of other strong modes can also be determined.

9.6 Standing Waves

If a spiral density wave is absorbed in the inner or outer regions, then spiral structure will be relatively short-lived. If, on the other hand, the wave reflects before it gets absorbed in the inner region, *standing waves* can be established. Then a *quasi-stationary pattern* can be set up in the disk, and spiral structure can last longer. Certain physical properties of the disk can help create standing wave patterns, also called *wave modes*.

Two basic mechanisms have been proposed for amplifying waves to prevent them from dying out. Sometimes these are called *WASERS* in analogy with lasers (that is, wave amplification by stimulated emission of radiation, where the radiation refers to the waves). In the WASER I mechanism, developed by James Mark, a short trailing wave moves toward the central region and is refracted in the vicinity of the bulge. Then it is converted into a long-wavelength trailing spiral that moves out. The long wave reaches corotation and is reflected back in at greater amplitude as a short trailing wave. The long, outward moving, leading wave also gets converted into a short-wavelength, outward-moving, trailing wave beyond corotation. This wave motion is shown in Figure 9.21.

The *swing amplifier* (also called the *WASER II* mechanism) was recognized by P. Goldreich and D. Lynden-Bell and developed further by Toomre. In this mechanism, a short leading spiral perturbation at corotation shears into a short trailing spiral over time due to differential rotation. The wave is amplified by self-gravity as it does this, because stars in the perturbation are brought closer together (see Figure 9.22). This amplification occurs at the expense of some disk rotational energy. In both WASER mechanisms, the waves are able to derive energy from the rotation of the disk.

For a wave to be quasi-stationary, or relatively long-lived, it must turn around in the central region and amplify at corotation. For example, there may be a reflection off the central bulge due to the high-velocity dispersion there (with no change in pitch angle) or refraction off the bulge (with a change in pitch angle). A reflection would cause short trailing waves to become short leading waves that move outward, and a refraction would cause short trailing waves to become long trailing waves that move out. Then the reflected or refracted wave would go back out to corotation and get amplified. Figure 9.23 shows an example of a wave reflection. The inner reflection is at the bulge, and the solid curves represent the pattern of the spiral. The dashed lines represent the outgoing reflected waves, which have about the same pitch angle as the incoming waves but with the opposite winding (that is, they are leading waves).

Where the outgoing wave intersects an incoming wave, an *interference pattern* will be created; the spiral arm brightness will be increased if wave crests

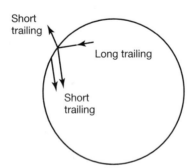

FIGURE 9.21   *Wave behavior at corotation, indicated by the circle. The long trailing wave is converted into inward- and outward-moving short trailing waves.*

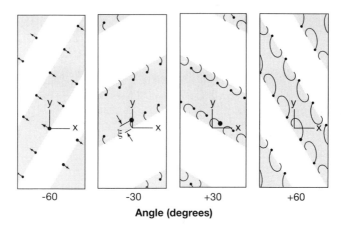

-60   -30   +30   +60

Angle (degrees)

FIGURE 9.22 *The principle of swing amplification (based on Toomre 1981.) The half-circles represent epicyclic motions, and the shaded pattern represents the wave. The angle of the wave swings from negative to positive in a differentially rotating system. In a disk, an outward leading spiral becomes an inward trailing spiral at corotation; the waves are reinforced there by energy from the disk rotation.*

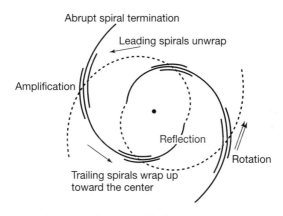

FIGURE 9.23 *Spiral wave reflection off of a central bulge results in an interference effect that causes incoming and outgoing waves to overlap. (From Elmegreen 1991.)*

meet and constructively interfere, or it will be diminished if wave crests and troughs meet and destructively interfere. This process probably accounts for the arm brightness variations in M81 (see Figure 5.13), where at one radius both arms suddenly diminish in brightness and then brighten at further radii.

Observations and models of M81 suggest the presence of a swing amplifier mechanism with a central reflection to create a mode, as illustrated in Figure

FIGURE 9.24   *Mode pattern in M81. (From Lowe et al. 1994.)*

9.24. This model may also apply to galaxies such as M101 (Figure 9.18), M100 (Figure 2.10), and M51 (Figure 2.2). The evidence for a mode is an interference pattern that shows up as gaps in the arm intensity on both main arms at the same distance from the center.

9.7 Stochastic Spiral Structure

On the basis of the observational evidence we discussed in Chapter 5, we infer that not all spiral galaxies have density waves. Differentially rotating disks can also generate transient spiral arm pieces in the gas and in the star-forming regions. Suppose that an OB association forms in a giant molecular cloud. The ionization pressures from the stars and shock waves from supernovae can trigger star formation in neighboring clouds. As the galaxy rotates, regions at slightly different radial distances are spread out by shear, with the inner star-forming regions moving at a higher angular rate than the nearby outer ones. Short spiral arm segments then develop and exist for as long as the individual stars survive, typically a few hundred million years. This means that a recurring stochastic spiral structure is created. Philip Seiden and Humberto Gerola developed a model that reproduces short-lived spiral structure in galaxies: the *stochastic self-propagating star formation* (SSPSF) theory. A typical model is shown in Figure 9.25. A feedback mechanism between star formation and death and the interstellar medium leads to a nearly exponential light profile in the disk, matching observations well for some flocculent galaxies and for the Sm class (Magellanic spirals). This model may also be important in describing the star formation bursts sometimes seen in irregular galaxies and especially dwarf irregular systems (see Chapter 11).

The stochastic model will rarely yield symmetric structure that appears to be coherent over long lengths, such as that observed in grand design galaxies, and it cannot produce spiral structure in the old Population I components. Hence,

FIGURE 9.25   *Three successive time intervals showing stochastically generated spiral structure. (From Gerola and Seiden 1978.)*

FIGURE 9.26   *Magnetic field lines follow the spiral arms in M51. (From Neininger 1992.)*

strong arm-interarm contrasts such as are seen in the near-infrared in grand design and multiple-arm structures are not explained by these models and instead require spiral density waves.

It should also be noted that early models for spiral structure considered magnetic fields to be the dominant force that organized the arms. Measurements of field strengths in spiral arms indicate that magnetic fields are too weak to organize these structures. Instead, the magnetic fields follow the gas as it moves in and out of the arms; see Figure 9.26.

Exercises

1. Determine the ratios of the resonance radii to R_{25} for ILR, inner 4:1, and corotation for the cases $\alpha = -0.2$, $\alpha = 0$, and $\alpha = 0.2$. Assume that the spiral arms end at R_{25}, which is identified as the OLR.

2. Determine the epicyclic frequency at $R = 0.6\,R_{25} = 7$ kpc for a galaxy with a flat rotation curve whose maximum velocity of 250 km s^{-1} is attained at $R = 0.2\,R_{25}$. If corotation is at $R = 0.5\,R_{25}$, what is this galaxy's pattern speed? Is $R = 0.6\,R_{25}$ a resonance radius?

3. A galaxy is observed to have a gap in the spiral arms at $R = 0.3\,R_{25}$ and an inner ring at $0.6\,R_{25}$. What is the slope of the rotation curve? Is there likely to be a WASER I mechanism, a WASER II mechanism, or neither? Why?

4. If dust lanes are observed to switch from the concave to the convex sides of spiral arms at $0.6\,R_{25}$ in a galaxy with a flat rotation curve, where is the inner 3:1 resonance?

5. For a galaxy with $M_B = -22$, calculate the epicyclic frequency at 5 kpc, assuming a peak velocity of 200 km s^{-1} at 3 kpc.

6. Assume that the outer Lindblad resonance is at $R_{25} = 20$ kpc for the galaxy in Exercise 2. What is the galaxy's pattern speed?

7. Consider two galaxies, with $M_B = -21$ and $M_B = -19$. Determine the ratio of the outer Lindblad resonance to the inner 4:1 resonance for each.

8. In Figure 9.10, determine the locations of OLR and corotation for a pattern speed of 20 km s^{-1} kpc^{-1}.

9. Fill in the algebraic steps to derive the radial velocity term in the rotating reference frame, shown on page 194.

Unsolved Problems

1. What is the mechanism that starts spiral density waves?
2. How do density waves affect star formation?

Useful Websites

See websites in Chapters 2, 4, and 7 for galaxy images and models.

Further Reading

Athanassoula, E. 1992. Morphology of bar orbits. *Monthly Notices of the Royal Astronomical Society* 259:328.

Berkhuijsen, E., C. Horellou, M. Krause, N. Neininger, A. Poezd, A. Shukurov, and D. Sokoloff. 1997. Magnetic fields in the disk and halo of M51. *Astronomy and Astrophysics* 318:700.

Bertin, G., and C. C. Lin. 1996. *Spiral structure in galaxies: A density wave theory*. Cambridge, MA: MIT Press.

Binney, J., and S. Tremaine. 1987. *Galactic dynamics*. Princeton: Princeton University Press.

Bowers, R., and T. Deeming. 1984. *Astrophysics II: Interstellar matter and galaxies*. Boston: Jones and Bartlett Publishers.

Buta, R. 1995. The catalogue of southern ringed galaxies. *Astrophysical Journal (Suppl.)* 96:39.

Elmegreen, B.G. 1991. Spiral types. In *Dynamics of galaxies and molecular cloud distributions*, ed. F. Combes and F. Casoli, 113. Dordrecht: Kluwer.

Elmegreen, B.G. 1992. Large-scale dynamics of the interstellar medium. In *The galactic interstellar medium: Saas-Fee advanced course 21*, ed. D. Pfenniger and P. Bartholdi. New York: Springer-Verlag.

Elmegreen, B.G., D.M. Elmegreen, and L. Montenegro. 1992. Optical tracers of spiral wave resonances in galaxies. II. Hidden three-arm spirals in a sample of 18 galaxies. *Astrophysical Journal (Suppl.)* 79:37.

Elmegreen, D.M., and B.G. Elmegreen. 1993. What puts the spiral in spiral galaxies? *Astronomy* 21:34.

Gerola, H., and P. Seiden. 1978. Stochastic star formation and spiral structure of galaxies. *Astrophysical Journal* 223:129.

Gottesman, S., and L. Weliachew. 1975. A high-resolution neutral-hydrogen study of the galaxy M81. *Astrophysical Journal* 195:23.

Kaufmann, D., and G. Contopoulos. 1996. Self-consistent models of barred spiral galaxies. *Astronomy and Astrophysics* 309:381.

Kormendy, J., and C. Norman. 1979. Observational constraints on driving mechanisms for spiral density waves. *Astrophysical Journal* 233:539.

Lin, C.C., and F. Shu. 1964. On the spiral structure of disk galaxies. *Astrophysical Journal* 140:646.

Lowe, S., W. Roberts, J. Yang, G. Bertin, and C. C. Lin. 1994. Modal approach to the morphology of spiral galaxies. 3. Application to the galaxy M81. *Astrophysical Journal* 427:184.

Neininger, N. 1992. The magnetic field structure of M51. *Astronomy and Astrophysics* 263:30.

Patsis, P., N. Hiotelis, G. Contopoulos, and P. Grosbol. 1994. Hydrodynamic simulations of open normal spirals: OLR, corotation, and 4/1 models. *Astronomy and Astrophysics* 286:46.

Rubin, V., N. Thonnard, W. Ford, and D. Burstein. 1982. Rotational properties of 23 Sb galaxies. *Astrophysical Journal* 261:439.

Toomre, A. 1981. What amplifies the spirals? In *The structure and evolution of normal galaxies*, ed. S.M. Fall and D. Lynden-Bell, 111. Cambridge: Cambridge University Press.

Visser, H. 1980. The dynamics of the spiral galaxy M81. I—Axisymmetric models and the stellar density wave. II—Gas dynamics and neutral hydrogen observations. *Astronomy and Astrophysics* 88:149.

CHAPTER **10**

Large-Scale Star Formation

Chapter Objectives: to understand star formation on a global scale and how to measure it

---

Toolbox:

star formation rate   Q parameter
Schmidt law   star formation efficiency
Jeans instability

10.1 Global Star Formation Rates and Histories

The interstellar medium provides a fundamental connection between disk dynamics and local physics, because it is there that new stars form and old stars recycle their ejected material. The *star formation rate* (SFR) is the number of stars formed per unit time. This rate should be normalized to the galaxy mass or gas mass when comparing different galaxies.

Measures of star formation activity include ultraviolet, Hα, and far-infrared fluxes, which all trace young high mass stars with ages $<10^8$ yr. Deriving the overall SFR from these fluxes requires an assumption about the initial mass function (IMF) and the mass/luminosity ratio for different stellar types as a function of their ages, as we considered in Chapter 4. Fluxes from uv and Hα only give a lower limit to star formation rates because of extinction. The conversion of Hα luminosities to total star formation rates for 0.1 to 100 M$_o$ stars, assuming a standard initial mass function, is

$$\frac{dM}{dt} = 7.07 \times 10^{-42} L_{H\alpha} \ M_o/yr$$

for star formation rate dM/dt in solar masses per year, and luminosity $L_{H\alpha}$ in solar luminosities. Emission at 2000 Å uv is also a useful indicator of the SFR, but not

216

many galaxies have been measured at this wavelength because it requires satellite observations.

The most widely available measures of star formation activity are far-infrared fluxes from the IRAS (Infrared Astronomical Satellite) survey. The global far-infrared luminosity observed in spiral galaxies is consistent with the number of high-mass OB stars required to ionize the gas. The IR flux is only an upper limit to the amount of star formation, because an estimated $1/2$ of the L_{FIR} comes from cirrus clouds. These are diffuse clouds that are far from high-mass stars; they contain dust heated by the general radiation field rather than nearby OB stars. On the basis of model fitting with a standard initial mass function, the conversion to SFR from infrared luminosity, including a factor of $1/2$ for cirrus emission, is given as the mass of stars formed per unit time, dM/dt:

$$\frac{dM}{dt} = 3.2 \times 10^{-10} L_{FIR} \ M_o/yr$$

for far-infrared luminosity L_{FIR} expressed in solar luminosities and the derived SFR expressed in solar masses per year.

In addition, Type II supernovae are indicators of massive star formation, because their progenitors are early-type stars; a measure of their rate per unit galaxy luminosity probes the star formation rate. Colors such as (U-B), (B-V), and (B-I) also give an indication of the average age of a region, because stellar populations redden as they evolve (see Chapter 4).

Blue luminosities indicate the average SFR over the last billion years in a galaxy. Assuming a constant SFR and a standard IMF, and ignoring post–main sequence stars, we can write the SFR in terms of blue luminosity as

$$\frac{dM}{dt} = 3.68 \times 10^{-10} L_B \ M_o/yr$$

for galaxy luminosity L_B in solar luminosities.

The overall star formation rate in normal spiral and irregular galaxies is typically ~1–5 $M_o \ yr^{-1}$, but it may be as low as $6 \times 10^{-4} \ M_o \ yr^{-1}$ in dwarf irregulars. Normalized to unit area, the average star formation rate is ~$5 \times 10^{-9} \ M_o \ yr^{-1} \ pc^{-2}$ in spirals and ~$2 \times 10^{-9} \ M_o \ yr^{-1} \ pc^{-2}$ for giant irregulars, with as much as a factor of 10 variation from galaxy to galaxy. The rates per unit area for the Milky Way, Large Magellanic Cloud, and Small Magellanic Cloud are similar to these.

The star formation rate per unit area is essentially constant in all galaxies of a given Hubble type. The fact that this rate is independent of Arm Class (see Chapter 5) indicates that spiral density waves do not cause a global increase in the star formation.

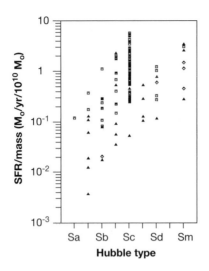

FIGURE 10.1   *The star formation rate (SFR) normalized to galaxy mass, as a function of Hubble type. (From Ferrini and Galli 1988.)*

The SFR per galaxy mass increases slightly with later Hubble type, as shown in Figure 10.1, because later types have relatively more gas than earlier types. This figure is based on SFR determined from Hα and uv luminosities.

The star formation rate per unit gas mass, measured from the ratio of L_{IR}/M_{H2}, is ~$5L_o/M_o$, independent of Hubble type. That is, even though the gas fraction varies as a function of Hubble type, the relative rate of conversion of gas into stars is approximately constant for all normal galaxies.

A comparison of the CO distribution, the blue light distribution (which measures the SFR over the last ~10^9 yr) and the Hα or IR distributions (which measure the current SFR for high-mass stars) yields the star formation history in a galaxy. The azimuthally averaged intensities of blue light, Hα, red continuum, and CO all tend to have the same radial distribution, whereas the H I distribution is essentially flat, as shown in Figure 10.2 for NGC 4254. The similarity between blue and Hα radial distributions suggests that normal spiral galaxies have had approximately constant star formation rates with radius over at least the last several million years.

Indeed, population synthesis models that match observations indicate that the overall star formation rates in spirals and large irregulars may have been approximately constant over the last billion years. In contrast, elliptical galaxies evidently had most of their star formation in the first billion years after their formation, as shown in Figure 10.3. This schematic diagram assumes that spiral bulges formed in the first billion years, after which disk formation began. The smoothness of the curves is not well established; as gas densities drop below the critical density, the star formation rates in the disk may change abruptly. In Chapter 11 we will consider galaxies undergoing bursts of star formation.

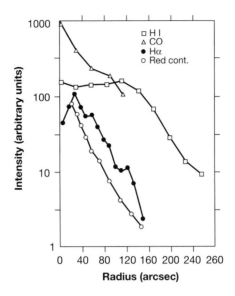

FIGURE 10.2 *Radial variation of gas and stars in NGC 4254. (From Kennicutt 1989.)*

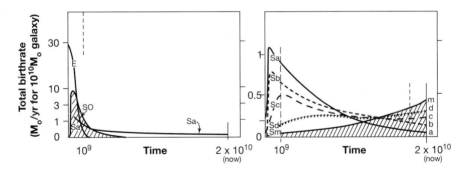

FIGURE 10.3 *Star formation rates as a function of time for different Hubble types. Ellipticals had most of their star formation within the first billion years, whereas spirals have had a steadier rate. (From Sandage 1986.)*

The rate at which the interstellar medium evolves in a galaxy is determined by the star formation rate in molecular clouds. If the star formation rate and gas supply M_{gas} are known, it is possible to compute the length of time for which the present SFR can be sustained. Of course, the evolution of the interstellar medium depends on the conversion of atomic to molecular gas and on the formation of stars of all masses. The present SFR can be measured only for high-mass stars, which we can represent as dM_{OBA}/dt. The *gas depletion time* is

$$t_{gas} = \frac{M_{gas}}{dM_{OBA}/dt}$$

If the present high-mass SFR is approximately constant with time, then the molecular interstellar medium will be cycled into high-mass stars (>5 M_o) on a time scale of several $\times\ 10^9$ yr, assuming a standard initial mass function. In turn, some of that material is returned to the interstellar medium via supernova remnants. Gas depletion then depends on the fraction of gas that is locked up in low-mass star formation, which is poorly determined. If the IMF includes more high-mass stars than usual, the gas supply will last longer owing to more rapid recycling from the high mass stars back into the interstellar medium. The estimated depletion times range from 10^9 to a few $\times\ 10^{10}$ yr in spiral and irregular galaxies, and up to 10^{12} yr in dwarf irregulars. These estimates are based on closed systems; galaxies that acquire new gas through infall of the intercloud medium (ICM) or through tidal interactions could form stars for a longer period.

10.2 Disk Instabilities

There is a correlation between the star formation rate per unit area and the gas density. The correlation has less dispersion if the gas density includes both atomic gas and molecular gas rather than just atomic or just molecular gas, as shown in Figure 10.4.

The *Schmidt law* relates the star formation rate per unit volume to a power of the gas density ρ:

$$\frac{SFR}{volume} \propto \rho^n$$

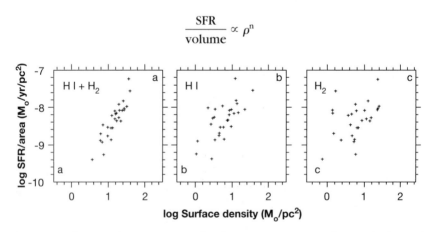

FIGURE 10.4   *Star formation rate (SFR) per unit area as a function of gas; the correlation is best when atomic and molecular gas are both included. (Figure from Buat et al. 1989.)*

The standard Schmidt law assumes that n = 2; observationally, n is usually from 1 to 2. This exponent may be explained from different points of view. For example, since the ratios of Hα/CO luminosity and blue/CO luminosity are nearly constant as a function of radius for most galaxies, the SFR for high mass stars and the long-term rate for intermediate mass stars might scale directly with the available supply of molecular gas. In this case, the power n is 1 for the molecular gas. However, if star formation is initialized by a gravitational instability, which has a time scale $\propto(G\rho)^{-0.5}$ (see below), then the star formation rate per unit mass depends on $\rho^{0.5}$ and the power in the Schmidt law is 1.5. Alternatively, if star formation occurs because of collisions between clouds, then n = 2 since the cloud interaction rate depends on the square of their density.

A region becomes gravitationally unstable when the gravitational potential energy exceeds the kinetic energy. From a dimensional analysis (that is, disregarding factors of π and 2, and so on), we have the condition

$$\frac{GMm}{r} > kT$$

for a cloud of total mass M, hydrogen mass m, radius r, and temperature T. As a cloud collapses, the gravitational energy is converted to kinetic energy, which must be dissipated through radiation or else the random motions of the particles will be too high for them to form a star.

Three-dimensional collapse processes are appropriate when considering small-scale star formation. Sir James Jeans derived the three-dimensional *Jeans instability* in 1902, which relates the length of an unstable region to a given density and temperature. Rewriting the total mass M of the region in terms of its density, we have $M = (4/3)\,\pi\,r^3\rho$. Then from a balance of gravitational and kinetic energies, the minimum length for instability is the *Jeans length* (L_J)

$$L_J \approx \sqrt{\frac{kT}{Gm\rho}}$$

A more exact analysis gives a value of 0.77 times this for a spherical cloud of initially uniform density (see references in Further Reading for more details). The *Jeans mass* M_J is the density times the volume contained within the Jeans length; for a sphere,

$$M_J = \frac{4\pi L_J^3 \rho}{3}$$

For large-scale star formation in a spiral galaxy disk, it is more appropriate to consider instabilities in two-dimensions. The volume density must be replaced by a surface density for the gas, Σ, which we can approximate from the gas mass M by

$$\Sigma \approx \frac{M}{r^2}$$

The kinetic energy can be written in terms of the disk velocity dispersion a. Then the Jeans length is given by

$$L_J = \frac{2a^2}{G\Sigma}$$

For large-scale gas instabilities in a rotating disk, V. Safronov showed that the epicyclic frequency κ must also be taken into account. Then there is a Q parameter that measures the degree of stability against perturbations,

$$Q = \frac{\kappa a}{\pi G \Sigma}$$

Toomre showed that for stars only, π in this equation should be replaced by 3.36. For $Q < 1$, the disk is unstable to radial perturbations or rings. Observationally, it is convenient to compare the gas surface density Σ with a critical surface density Σ_c. The critical gas surface density can be written as

$$\Sigma_c = \frac{0.7\kappa a}{\pi G}$$

Thus, the Q parameter measures the approximate ratio of gas surface density to critical surface density. Empirically, star formation occurs when $\Sigma/\Sigma_c > 1$. Figure 10.5 shows the observed surface density relative to the critical density in some late-type galaxies and the Milky Way. The outermost observed H II regions occur where the surface density drops below the critical value. Typical surface densities required for star formation in a disk are on the order of 1 to 5 M_o pc^{-2}, depending on κ.

Irregular galaxies often have extensive H I distributions. However, their gas column density is less than the critical column density beyond the Holmberg

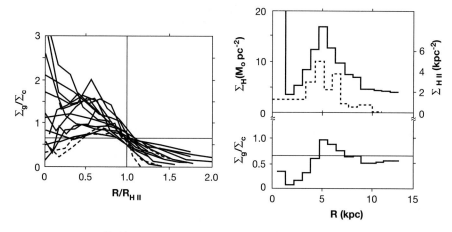

FIGURE 10.5   (left) The ratio of the gas density to the critical density for several galaxies as a function of distance, scaled to the radius of the outermost H II region. (right) The ratio of the gas density to the critical density in the Milky Way (bottom) and the radial distributions of H I and H₂ (solid line) and H II (dashed line) (top). (Figures from Kennicutt 1989.)

radius. In S0 galaxies, the density is less than the critical density everywhere in the disk.

Similar arguments can be applied to star formation as a result of ring instabilities. For example, in NGC 1300 (Figure 5.9), the outer ring has a gas column density greater than the critical value, so there is star formation in the ring. In contrast, regions inside and outside the ring are stable against collapse. In the barred spiral galaxy NGC 3351 (Figure 9.15), a circumnuclear ring has star formation in regions of high gas density where Q is less than the threshold.

By dimensional analysis, we can get the dependency of a gravitational collapse time on the initial density. If the acceleration of the collapse were constant and the density remained uniform throughout a collapsing sphere, then the acceleration g(r) for a three-dimensional instability would be given by

$$g(r) = \frac{GM(r)}{r^2} = \frac{4}{3\pi G \rho r}$$

The free-fall collapse time would be

$$t_{ff} = \sqrt{\frac{2r}{g(r)}} = \sqrt{\frac{3}{2\pi G \rho}}$$

Density varies with radius in a real collapsing cloud, so the constants change; in general, the time scale for a gravitational instability is

$$t_{ff} \propto \frac{1}{\sqrt{G\rho}}$$

A more detailed analysis of Jeans length and instability time scales is given in Appendix 4.

10.3 Complexes and Propagating Star Formation

The largest aggregates of star-forming regions in a galaxy are star complexes, with sizes up to about 1 kpc; an example is shown in Figure 10.6.

The diameters of the largest complexes, D_c, scale with the galaxy luminosity, L_B:

$$D_c \propto L_B^{0.4}$$

for a range spanning a factor of 10^4 in galaxy luminosity, as shown in Figure 10.7. The size of the largest complex in a galaxy is independent of the Arm Class or bar type. The correlation between complex size and galaxy luminosity

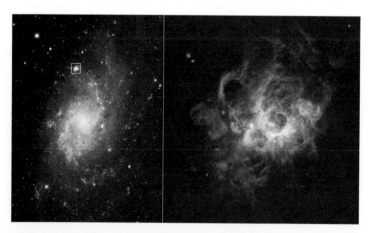

FIGURE 10.6   *M33 (left) and an enlargement (right) of the star-forming complex NGC 604, which is circled in the left-hand image. (Imaged by H. Yang, University of Illinois; and NASA using the Hubble Space Telescope WFPC2.)*

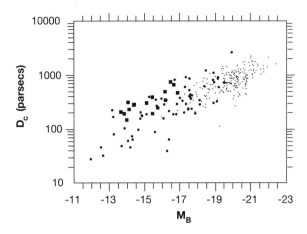

FIGURE 10.7   *Largest complex diameter (D$_c$) as a function of galaxy absolute magnitude. Brighter galaxies are spirals, and fainter galaxies are dwarf irregulars; the squares are Blue Compact Dwarfs. (From Elmegreen et al. 1996.)*

can be interpreted in terms of a large-scale disk instability. Complex sizes are slightly larger relative to the galaxy size in dwarf galaxies compared with spiral galaxies (by a factor of 3).

Star formation in giant molecular clouds includes high mass stars that ionize the gas around them. Radiation pressure and stellar winds from OB stars and associations, and shocks from their subsequent supernovae, lead to new generations of star formation in nearby gas. Often over scales of several hundred parsecs, there is a sequence of protostars, H II regions, and older clusters. This progression is a result of *propagating star formation*. An example is shown in Figure 10.8, where several generations of stars have formed along a complex. Star formation that is initiated by external pressures is sometimes called *stimulated star formation*. In contrast, smaller, more isolated regions such as Bok globules (see Chapter 6) may have *spontaneous star formation* in which the gas is gravitationally unstable to collapse even in the absence of obvious external pressure.

Measurements of ages in complexes have been made by examining the distributions of high-mass stars, supergiants, Cepheids, Wolf-Rayet stars, masers, and H II regions of our Galaxy and other nearby spirals. Age gradients observed in the complexes are typically 0.3×10^7 yr to 1.2×10^7 yr per 300–500 pc. One interpretation of these age spreads is that they are regions of propagating star formation; typical gas motions of 10 km s^{-1} amount to 100 pc in 10^7 yr, which would account for the gradients. An example of an age gradient is shown in Figure 10.9 for a complex in our Galaxy.

The Lindblad expanding ring in our Galaxy, which is part of Gould's Belt (see Chapter 8), may be an example of propagating star formation occurring in a shell of material distributed around the source of the pressure. In this scenario, the expanding ring is the result of star formation that may have been triggered by

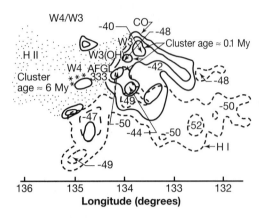

FIGURE 10.8 *W3/W4 in the Milky Way, as traced by CO temperature contours (solid lines) and H I (dashed lines). The velocities of H I are indicated in each region. (From Elmegreen and Wang 1988.)*

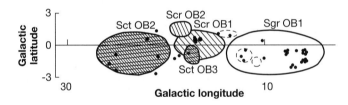

FIGURE 10.9   *Age gradient across the star-gas complex SGC2 in the Milky Way. Dots are high-mass main sequence stars. Cross-hatched regions are 7.2×10^6 yr; single-hatched regions are $5–6.5 \times 10^6$ yr; white regions are 4.4×10^6 yr. The age decreases from left to right. (From Sitnik 1991.)*

impacts on the Galactic plane from high-velocity clouds (see Chapter 6). A schematic diagram of the locations of star-forming cloud complexes around the ring is shown in Figure 10.10.

The resulting supershell from this type of event appears to be common in many galaxies, as hypothesized by G. Tenorio-Tagle and P. Bodenheimer. Supershell sizes typically range from 100 pc to 1 kpc and require energies of 10^{49}–10^{52} erg. A schematic diagram of the supershell region around a superbubble is shown in Figure 10.11.

10.4 H II Regions

H II regions have a wide range of sizes, depending on the central star temperature and local gas density (see Chapter 6). In galaxies of the same luminosity, the diameters of the largest H II regions are larger in later-type spirals than in early-

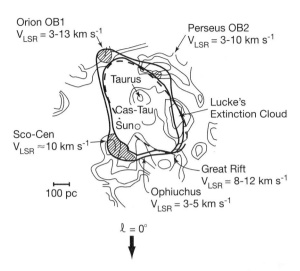

FIGURE 10.10 *Lindblad's expanding ring (indicated by the central oval), showing cloud complexes and their velocities. (From B. Elmegreen 1992.)*

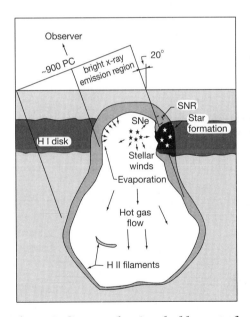

FIGURE 10.11 *Schematic diagram showing the blowout of a superbubble extending beyond the H I scale height. (From Wang and Helfand 1991.)*

type spirals. The *cumulative luminosity function* is the number (N) of H II regions brighter than a given luminosity as a function of luminosity. The observed function is similar for early- and late-type systems, as shown in Figure 10.12.

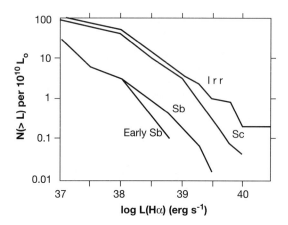

FIGURE 10.12   *Cumulative H II region luminosity function versus luminosity for different galaxy types. (Based on Kennicutt et al. 1989.)*

The positional distribution of all H II regions in a galaxy shows little global structure. However, when H II regions are sorted by size or luminosity, the spiral pattern is discernible. Giant H II regions (with luminosities $>10^{38}$ erg s^{-1}) are concentrated in spiral arms, but smaller H II regions are more uniformly distributed throughout the disk, as shown in Figure 10.13. Giant H I clouds, giant molecular clouds, and giant H II regions all have similar spatial distributions and are prominent in spiral arms, sometimes with regular spacing (see Figure 10.14). The spacing and regularity are consistent with a large-scale gravitational collapse.

Based on observations of several galaxies, the number of large H II regions in arms is two to five times the number in the interarms. This result could be from the collection of more material in the arm, from excess *triggering* of star formation in existing clouds by a density wave, or from both. Triggering means that the

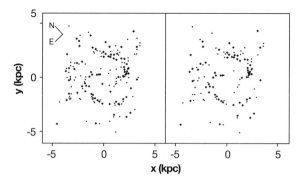

FIGURE 10.13   *All H II regions in M83 (left); giant H II regions in M83 (right). Note the more prominent spiral pattern in the right-hand image. (From Rumstay and Kaufman 1983.)*

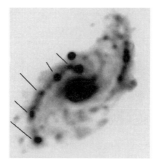

FIGURE 10.14 *"Beads on a string" (indicated by arrows) in the central spiral arms of NGC 5248. (Image from the STScI Digital Sky Survey.)*

star formation rate per unit gas mass is larger than it would have been in the absence of a density wave. (Refer to Appendix 4 for a brief discussion of shock compression.) If the gas is merely collected into the arms, then the luminosity function of H II regions should be displaced upward relative to the interarm luminosity function, so there should be more H II regions at all luminosities. If each cloud forms more stars than it would have in the interarm, then the luminosity curve will be shifted horizontally. If small clouds systematically collect together to make large clouds and therefore large H II regions, with a corresponding deficit in small clouds and small H II regions, then the luminosity function will be shallower in the arms than in the interarms, regardless of any shift. All of these luminosity functions are assumed to have a cutoff at the same low luminosity level because of observational limitations. The three possibilities are shown schematically in Figure 10.15.

Observed luminosity functions for arm and interarm regions in several late-type galaxies are shown in Figure 10.16. Because triggering and nontriggering show a similar shift in the H II region luminosity function, these data do not clearly indicate what is happening in real galaxies.

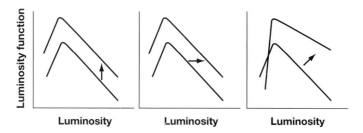

FIGURE 10.15 *(left) An upward-shifted luminosity function results from an increase in the amount of star formation. (right) A rightward-shifted luminosity function results from an increase in the number of high-luminosity objects.*

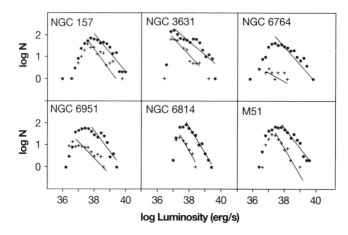

FIGURE 10.16 *Luminosity function slopes of H II regions in the arms (top points in each graph) and interarms (bottom points) of six late-type spirals. The best-fit slopes are shown; they indicate a lack of significant triggering. (From Rozas et al. 1996.)*

10.5 Azimuthal Variations

Azimuthal variations are apparent in CO, H I, Hα, and optical light, which reflect the presence of spiral arms. Some observations indicate that the surface density ratio of H_2/H I is highest on spiral arms, with a factor <10 enhancement relative to the ratio in the interarm. This result suggests enhanced molecular cloud formation efficiency in the arms.

In general, there are dust lanes on the concave sides of spiral arms inside corotation, with H II regions and older stars displaced downstream from the dust lanes. In galaxies observed with high resolution, such as M31, the peak CO emission and the largest clumps of H I emission are in the spiral arms, whereas small clouds exist throughout the disk. These results are similar to the spatial distributions that we see in our Galaxy. Some galaxies have H I, CO, and Hα peaks that are coincident, and typically a few hundred parsecs downstream from the dust lanes. Hβ peaks tend to be anticoincident with H I, because Hβ can be seen only where there is low extinction. Figure 10.17 shows observations of a spiral arm in M83.

In other galaxies, such as M51 and M33, the H I is displaced 50–100 pc downstream from the CO. The distribution of CO and Hα in M51 is shown in Figure 10.18. The CO and H I emissions closely follow the optical arms. In M81, the CO emission is weak, and the H II and H I peaks are downstream from the dust lane. The details of gas distributions depend on cloud formation and destruction as well as on star formation.

Intensity peaks have been observed at optical and radio wavelengths along the arms of several galaxies. Often the peaks are greater for younger stars than for older stars. Determining the extent of any triggering requires accurate interarm gas and star measurements for comparison, but these are often not well observed.

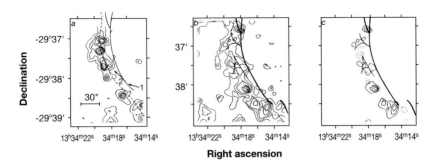

FIGURE 10.17 *Observations of an arm in M83 in Hβ (left), with the H I ridge represented by a line running through the contours, H I (middle), and a combination of Hβ (dots) and H I (contours) (right). The thick line represents the dust lane. The horizontal bar on the left corresponds to 530 pc for a distance to M83 of 3.7 Mpc. (From Allen et al. 1986.)*

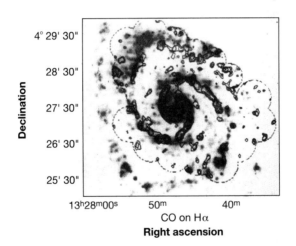

FIGURE 10.18 *CO contours overlaid on Hα grayscale in M51. (From Rand and Kulkarni 1990.)*

Color gradients have been observed across arms in NGC 2903 (see Figure 10.19). Asymmetric color gradients are expected from the progression of stellar evolution through a density wave, but they can also arise from extinction at the inner edge of the arm.

10.6 Star Formation Efficiencies

Star formation may be measured in terms of a cloud's efficiency rather than its rate of production of young stars. The *star formation efficiency* (SFE) can be de-

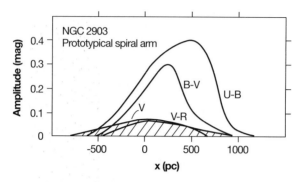

FIGURE 10.19 *Asymmetric color profiles across the arms of NGC 2903. (From Hodge et al. 1990.)*

fined to be the mass of stars divided by the sum of the total stellar mass and to-tal cloud mass:

$$SFE = \frac{M_{*tot}}{M_{*tot} + M_{gas\,tot}}$$

where M_{*tot} is the total star mass and $M_{gas\,tot}$ is the total gas mass. The SFE in a young region, based on optical observations, usually amounts to only a few per-cent overall; because many embedded stars are missed, infrared observations of embedded stars sometimes increases the SFE to 10%–30%. One mechanism that may limit the efficiency of star formation, at least for massive stars, is the de-struction of star-forming clouds by ionization.

An integration of the SFR over time gives the total mass of stars formed. This quantity is not straightforward to derive, because high- and low-mass stars evolve at different rates, and extinction obscures low-mass stars more than high-mass stars. Furthermore, the initial gas content is uncertain because high-mass stars can destroy the gas clouds in which they formed. The ratio of star to gas mass can be measured by either masses or luminosities, so in the literature, efficiency has been measured by L_*/L_{gas}, L_*/M_{gas}, M_*/M_{gas}, and the ratio of Σ_*/N for surface brightness di-vided by gas column density. Sometimes the SFE is defined differently, as the star formation rate per unit mass, SFR/M_{gas}, which we discussed in Section 10.1.

In our Galaxy, the nearby star-forming complexes in Taurus (140 pc distant), Ophiuchus (160 pc), and Orion (500 pc) have been mapped in detail. Each con-tains a gas mass of about $10^4–10^5\ M_\odot$; Orion is the most massive. The spatial dis-tribution of star formation varies in these clouds. In Taurus, stars form individu-ally and in small groups throughout the cloud. In contrast, Ophiuchus and Orion stars are concentrated in a small number (1–3) of dense clusters. The Orion neb-ula is shown in Figure 10.20.

On the basis of infrared observations of Taurus-Auriga, there are 100–140 main sequence stars of $L > 0.5L_\odot$, with a median mass ~0.6 $M_\odot$, so $M_{*tot} \sim 70$. This gives a star formation efficiency of $70/10^4 = 0.7\%$. In ρ Ophiuchus, $M_{*tot} \sim 82$ with

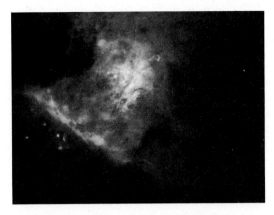

FIGURE 10.20 *The Orion nebula. (Imaged with Hubble Space Telescope WFPC2 by C. R. O'Dell and S. K. Wong, Rice University; and NASA.)*

SFE = 0.8%. Orion A has ~600 stars with a median mass of ~1.0 M_o, so 600 M_o/10^5 M_o ~0.6%. In Orion B, the SFE is ~3%–4%. Orion B has 5 cores >200 M_o, but only three cores are producing most of the young stellar objects; the SFE ranges from a high of 20%–40% to a low of ~7%. In each cloud, at most a few percent of the cloud's mass goes into making stars.

In comparing different regions, it is important to note that the scale over which star formation is measured yields different star formation efficiencies. Obviously, if the scale includes only the star-forming region, the efficiency will be higher than if an extended region is included. The star formation efficiency for nearby clouds in our Galaxy linearly decreases with increasing size, as shown in Figure 10.21.

In Chapter 11, we will consider central star formation and enhanced star formation activity, as related to tidal interactions of galaxies, bar-driven gas flow, and episodic bursts.

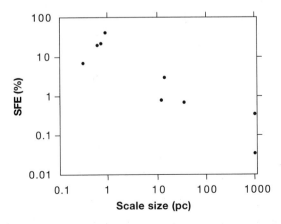

FIGURE 10.21 *Star formation efficiency (SFE) decreases as the size of the measured region increases. (From Elmegreen 1992.)*

Exercises

1. Suppose that the velocity dispersion in a galaxy disk is 15 km s^{-1}, with a surface density of 3 M_o/pc^2. Determine the disk stability.
2. Suppose that a $10^{4.5}$ M_o giant molecular cloud in our galaxy has an observed far-infrared luminosity of 10^6 L_o. What is the high-mass star formation rate? If this rate is maintained over a total of 10^7 yr, what is the star formation efficiency of the cloud?
3. Calculate the gas depletion time if star formation continues at a rate of 2 M_o yr^{-1} for a disk gas density of 2 cm^{-3}.
4. Determine the Jeans length in a molecular cloud with a temperature of 10 K and a density of 100 cm^{-3}.
5. Determine the Q parameter in a galaxy with a flat rotation curve, a peak velocity of 200 km s^{-1}, and a disk surface density of 5 $\times$ 10^{20} cm^{-2}.
6. (Refer to Appendix 4.) Compare the compression in an adiabatic shock and an isothermal shock for Mach 5.

Unsolved Problems

1. How do stars form?
2. How does star formation vary in different galaxies and in different parts of a galaxy?

Useful Websites

Recent mm observations of galaxies with BIMA (Berkeley-Illinois-Maryland Array) are posted on http://astron.berkeley.edu

Further Reading

Allen, R., P. Atherton, and R. Tilanus. 1986. Large-scale dissociation of molecular gas in galaxies by newly formed stars. *Nature* 319:296.

Buat, V., J. Deharveng, and J. Donas. 1989. Star formation rate and gas surface density in late-type galaxies. *Astronomy and Astrophysics* 223:42.

Cepa, J., and J. Beckman. 1990. Diagnostics of the underlying physical differences between flocculent and well-defined spiral galaxies. *Astronomy and Astrophysics* 239:85.

Efremov, Y. 1995. Star complexes and associations: Fundamental and elementary cells of star formation. *Astronomical Journal* 110:2757.

Elmegreen, B.G. 1989. Molecular cloud formation by gravitational instabilities in a clumpy interstellar medium. *Astrophysical Journal* 344:306.

Elmegreen, B.G. 1992. Triggered Star Formation. In *Star formation in stellar systems, III Canarie Islands Winter School*, ed. G. Tenorio-Tagle, M. Prieto, and F. Sanchez, 381. Cambridge: Cambridge University Press.

Elmegreen, B.G., and D.M. Elmegreen. 1986. Do density waves trigger star formation? *Astrophysical Journal* 311:554.

Elmegreen, B.G., D.M. Elmegreen, J. Salzer, and H. Mann. 1996. On the size and formation mechanism of star complexes in Sm, Im, and BCD galaxies. *Astrophysical Journal* 467:579.

Elmegreen, B.G., and A. Parravano. 1994. When star formation stops: Galaxy edges and low surface brightness disks. *Astrophysical Journal Letters* 435:121.

Elmegreen, B., and M. Wang. 1988. Triggering mechanisms for star formation. In *Molecular clouds in the Milky Way and external galaxies*, ed. R. Dickman, R. Snell, and J. Young, 240. New York: Springer-Verlag.

Elmegreen, D.M. 1992. Distribution and triggering of star-forming regions. In *Star-forming galaxies and their interstellar media*, ed. J. Franco, F. Ferrini, and G. Tenorio-Tagle, 106. Cambridge: Cambridge University Press.

Ferrini, F., and D. Galli. 1988. A model of spiral-galaxy evolution. I. Galaxy morphology and star formation rate. *Astronomy and Astrophysics* 195:27.

Herbst, W., and G. Assousa. 1977. Observational evidence for supernova-induced star formation—Canis Major R1. *Astrophysical Journal* 217:473.

Hodge, P., E. Jaderlund, and M. Meakes. 1990. UBVR CCD photometry of the spiral galaxy NGC 2903. *Publications of the Astronomical Society of the Pacific* 102:1263.

Hunter, D., and J. Gallagher. 1985. Star-forming properties and histories of dwarf irregular galaxies. Down but not out. *Astrophysical Journal (Suppl.)* 58:533.

Hunter, D., and J. Gallagher. 1986. Stellar populations and star formation in irregular galaxies. *Publications of the Astronomical Society of the Pacific* 98:5.

Kennicutt, R. 1989. The star formation law in galactic disks. *Astrophysical Journal* 344:685.

Kennicutt, R. 1992. The history of star formation in galaxies. In *Star formation in stellar systems, III Canarie Islands Winter School*, ed. G. Tenorio-Tagle, M. Prieto, and F. Sanchez, 191. Cambridge: Cambridge University Press.

Kennicutt, R., B. Edgar, and P. Hodge. 1989. Properties of H II region populations in galaxies. II. The H II region luminosity function. *Astrophysical Journal* 337:761.

Knapen, J., N. Arnth-Jensen, J. Cepa, and J. Beckman. 1993. Statistics and properties of H II regions in NGC 6814. *Astronomical Journal* 106:56.

Lada, E., K. Strom, and P. Myers. 1991. Environments of star formation: relationship between molecular clouds, dense cores, and young stars. In *Protostars and planets III*, ed. E. Levy, J. Lunine, and M. Matthews. Tucson: University of Arizona Press.

Lord, S., and J. Kenney. 1991. A molecular ridge of gas offset from the dust lane in a spiral arm of M83. *Astrophysical Journal* 381:130.

Rand, R. 1992. The H II region luminosity function of M51. *Astronomical Journal* 103:815.

Rand, R., and S. Kulkarni. 1990. M51: Molecular spiral arms, giant molecular associations, and superclouds. *Astrophysical Journal* 349:L43.

Rozas, M., J. Beckman, and J. Knapen. 1996. Statistics and properties of H II regions in a sample of grand design galaxies. I. Luminosity functions. *Astronomy and Astrophysics* 307:735.

Rumstay, K., and M. Kaufman. 1983. H II regions and star formation in M83 and M33. *Astrophysical Journal* 274:611.

Safronov, V. 1960. On the gravitational instability in flattened systems with axial symmetry and nonuniform rotation. *Annal d'Astrophysique* 23:901.

Sandage, A. 1986. Star formation rates, galaxy morphology, and the Hubble sequence. *Astronomy and Astrophysics* 161:89.

Scoville, N. and J.S. Young. 1991. Molecular gas in galaxies. *Annual Review of Astronomy and Astrophysics* 29:581.

Sitnik, T. 1991. Age distribution of stars and groups of stars in large-scale galactic complexes of stars and gas. *Soviet Astronomy Letters* 17:1.

Tenorio-Tagle, G., and P. Bodenheimer. 1988. Large-scale expanding superstructures in galaxies. *Annual Review of Astronomy and Astrophysics* 26:145.

Toomre, A. 1964. On the gravitational stability of a disk of stars. *Astrophysical Journal* 139:1217.

Vogel, S., S. Kulkarni, and N. Scoville. 1988. Star formation in giant molecular associations synchronized by a spiral density wave. *Nature* 334:402.

Wang, Q., and D. Helfand. 1991. LMC-2 as the blowout of a hot superbubble. *Astrophysical Journal* 379:327.

Young, J., L. Allen, J. Kenney, A. Hesser, and B. Rownd. 1996. The global rate and efficiency of star formation in spiral galaxies as a function of morphology and environment. *Astronomical Journal* 112:1903.

CHAPTER **11**
Starbursts and Active Galaxies

Chapter Objectives: to understand high levels of star formation and black hole accretion disks

---

Toolbox:

starbursts
active galactic nuclei
black holes

synchrotron emission
unified models

11.1 Starburst Activity

Galaxies with unusually high levels of star formation are referred to as *starburst* galaxies. They were first identified by their strong optical emission lines. The Hα emission line is especially prominent in active star-forming galaxies, with an equivalent width as much as 10 times higher than in normal spirals of the same Hubble type. Other emission line strengths depend on the nature of the star-forming region and the galaxy's metallicity. For example, strong [N II]/Hα and [S II]/Hα lines are characteristic of shocked gas, while strong broad lines of [O II], [O III], [N II], and [S II] are indicative of very luminous supernova remnants. Magellanic irregular starbursts have higher excitations because they are metal-poor.

Starburst spectra in the infrared (1–1000 μ) resemble those of compact H II regions in our galaxy and also include infrared contributions from diffuse emission. Ultraluminous ($L_{IR} > 10^{12}$ L_o) and starburst galaxies have high infrared surface brightnesses of $\sim 10^5$–10^7 L_o pc^{-2}, compared with the Milky Way disk value of ~ 10 L_o pc^{-2}. The infrared excess is interpreted to result from star formation; IRAS observations are very useful in probing such activity.

Regions of enhanced star formation activity may occur in extended tidal arms, main disk spiral arms, inner rings, or nuclear regions, depending on the details of the gas distribution and dynamics. In integrated spectra, it is difficult to distinguish between global and nuclear starbursting. However, in galaxies with global bursting, the overall blue continuum is much stronger than in galaxies with nuclear

237

bursting. Star formation rates increase as densities increase, as we saw in Chapter 10. Most massive star formation in starbursts occurs in regions of high optical depth. We will first consider standard starburst galaxies with central star formation enhancements and then consider enhanced star formation in disks.

High L/M ratios observed in the centers of starburst galaxies imply either the formation of predominantly high-mass stars or an increased efficiency overall. The starburst region may be local or may dominate the disk; for example, the overall ratios of infrared luminosity to total molecular gas, L_{IR}/total H_2, range from only 4 L_o/M_o (similar to the Milky Way disk) in the disks of nuclear starburst galaxies to 200 L_o/M_o in dwarf starburst galaxies (e.g., Mk 231).

As an example of a starburst galaxy, consider the prototype M82 (see Figure 2.17), a companion of M81. It is a Type II irregular, with a light/mass ratio of ~20 L_o/M_o overall, and nearly 200 L_o/M_o in the central region. Molecular distributions show elongations, plus a double peak that implies an edge-on torus with a radius of 200 pc. The total gas mass in this structure is ~10^8 M_o. In some regions, the high gas density of n ~ 5×10^4 cm^{-3} allows detection of weaker molecules such as HCN, CS, and HCO$^+$.

There is a central peak at 2μ, which is interpreted to be a nuclear star cluster with supergiant stars produced from a starburst. The implied star formation rate based on Lyman continuum photons is ~10 M_o yr^{-1}, assuming a standard IMF. At this rate, the gas in a 200 pc torus would be converted into stars in 10^7 yr. To maintain its starburst, it would have to have a mass inflow, such as a bar would cause; alternatively, the burst could be short-lived. Figure 11.1 is a schematic diagram of the process by which high-mass stars form and lead to x-ray emission. Such a process may be the fueling mechanism for the accretion disk of a central black hole as well (see Section 11.5).

In contrast to M82, NGC 2903 (Figure 11.2) is a spiral galaxy with a normal disk appearance. It has a central bar and a flocculent inner structure with two long symmetric outer arms. Its nuclear starburst is recognized by *hotspots* with

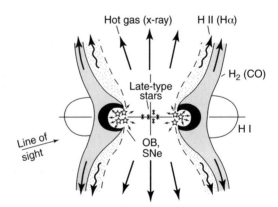

FIGURE 11.1 *Schematic view of the starburst nucleus of M82. (From Tomisaka and Bregman 1993.)*

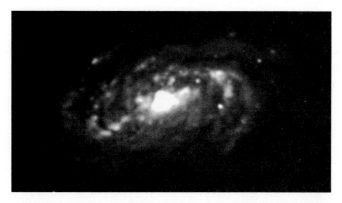

FIGURE 11.2 *The central regions are shown in B band for NGC 2903, an SAB galaxy with a starburst nucleus but an overall normal M/L ratio. (Imaged with the Burrell-Schmidt telescope at KPNO by the author.)*

strong ultraviolet light and H II region lines near the center; however, its nuclear molecular mass and mass surface density are approximately the same as in the Milky Way. Its OB luminosity is ~2.1 × 10^9 L_o in the central 8 arcsec, assuming a standard IMF. The mass of H_2 in the central 21 arcsec is 2.9 × 10^8 M_o. NGC 2903 has an overall ratio of L_{FIR}/M_{H_2} = 2.2 L_o/M_o, comparable to the ratio in the disks of normal spirals. If the gas is distributed over the central 8 arcsec only, then the L_{FIR}/M_{H_2} ratio is ~7 L_o/M_o, just a few times normal. This result implies that a nuclear starburst does not necessarily have an exceptionally high star formation efficiency but may result just from a large amount of molecular fuel forced into a small nuclear region. Recent Hubble Space Telescope observations are being used to determine the light/mass ratio, which could be much higher than 7 L_o/M_o if the emitting region is smaller than 8 arcsec.

Observations with the Hubble Space Telescope have revealed the presence of young globular clusters in many starburst galaxies. Often there are neighboring galaxies that are implicated in the bursting. However, young massive globular star clusters are also observed in starburst galaxies such as NGC 253 (Figure 11.3), which is not obviously interacting. Recent theoretical work has shown that extemely dense and massive clusters are able to form in high-pressure turbulent gas, such as that observed in the regions of the young globular clusters.

Some ultraluminous galaxies may have high luminosities from merging, with radiation from both energetic starburst activity and a nonthermal accretion disk. High resolution HST images reveal that the centers of many nearby elliptical galaxies are also undergoing intense star formation, presumably due to accretion of gas from small neighbors, with subsequent star formation. Some elliptical galaxies contain two or more nuclei; when these galaxies also have other evidence for interactions, such as excess light in their outer parts, then merging is inferred. In galaxies with an absence of obvious merger remnants, a model with a black hole emerges as a likely candidate to account for double nuclei. For example, in galaxies such as M31, models of the observed double

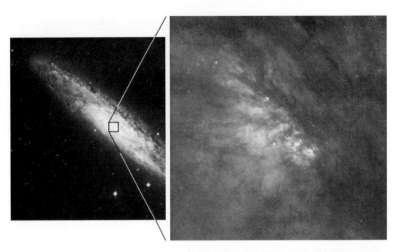

FIGURE 11.3   *NGC 253. The central region (right) shows a small bar. (Imaged with Hubble Space Telescope WFPC2 by J. Gallagher, University of Wisconsin; A. Watson, Lowell Observatory; and NASA.)*

nucleus (see Figure 11.4) indicate a massive cluster orbiting a central black hole in an eccentric orbit.

Central starbursting implies that there must be some mechanism that triggers excess star formation in the nucleus. Bar-driven gas infall is one mechanism for generating high gas concentrations. For example, bars are correlated with binary galaxies. The tidal force of a perturbing galaxy may induce infall of gas to nuclear regions because of a loss of angular momentum; n-body simulations of these actions are consistent with observations. The bar (which may form as a result of the interaction) then removes kinetic energy and angular momentum through shocking of the gas around it. CO observations reveal that gas is infalling along bars, as expected from theory. According to models of observations, a typical infall rate is $\sim 1 \ M_o \ yr^{-1}$.

Galaxies with bars often have rings, because bars produce torques on the gas near them. Gas outside corotation moves outward, while gas inside corotation

FIGURE 11.4   *M31: Double nucleus (left), ground-based view (middle), core (right). (Core observed with Hubble Space Telescope WF/PC by T. Lauer, NOAO; and NASA.)*

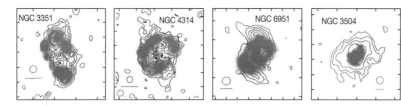

FIGURE 11.5 *CO contours on Hα grayscale maps show a variety of morphologies in the central regions of barred galaxies. NGC 3351 has twin peaks of emission; NGC 4314 shows a molecular ring; NGC 6951 has molecular spiral arms; and NGC 3504 has a central peak. (From Kenney 1996.)*

moves inward. Gas also accumulates at the Inner Lindblad Resonance (ILR). In galaxies without an Inner Lindblad Resonance, there is nothing to prevent the gas from going into the central regions. In these cases, a burst of star formation occurs near the nucleus, as shown for NGC 3504 and NGC 6951 in Figure 11.5. If there are two ILRs, gas collects in an inner ring. The piling up of material in the ILR region leads to an increased gas density in excess of the critical density for gravitational collapse, so stars form. CO contours and Hα hotspots are shown in Figure 11.5 for the hotspot galaxies NGC 3351 and NGC 4314 (see also Figure 9.15).

11.2 Quiescent and Starburst Dwarf Galaxies

Blue compact dwarf galaxies (BCDs) have a bright ultraviolet continuum, presumed to result from starburst activity occurring now (see Figure 4.16 for an example). Markarian first cataloged such objects on the basis of objective prism surveys. An evolutionary model has been proposed in which low surface brightness (LSB) dwarfs represent the quiescent phase and BCDs the starburst phase of the same object. This scenario is based on an interpretation of the galaxy spectra as well as slight bumps in the stellar mass functions, as described further in Section 11.4. BCD galaxies are about 2 magnitudes brighter than their corresponding quiescent phase. The starburst phase is transient in BCDs, because the gas depletion time is only 10^8–10^9 years. The galaxy becomes very blue for about 10^7 years, due to high mass star formation, and then reddens as the stars evolve to supergiants. Although the evolutionary process is not well understood, the stochastic star formation model (see Chapter 9) accounts for the episodic bursting nature of such galaxies in terms of their small mass.

The median H I content is higher in LSB dwarf galaxies than in BCD galaxies. Low surface brightness dwarfs have a factor of about 10 less infrared emission than BCDs: 10^7 to 10^8 L_o compared with 10^8 to 10^9 L_o. BCDs have an infrared luminosity per unit H I mass that is essentially the same as in normal spirals. However, the ratio of IR to B luminosity, L_{IR}/L_B, has a higher range in BCDs than in spirals and irregulars: ~2 to 15 for BCDs, ~0.2 to 5 for spirals. The higher ratio indicates a higher star formation rate in the BCDs at present.

Objective prism surveys have identified many active star-forming galaxies on the basis of strong emission lines or unusually high levels of ultraviolet emission. These *emission line galaxies*, many of which are BCDs, tend to be less clustered (see Section 12.1) than galaxies without emission lines.

11.3 Star Formation in Interacting Galaxies

Collisions of gas-rich galaxies are one cause of spectacularly luminous starbursts. Interactions between galaxies can generate tidal arms that lead to compression and increased random motions in the gas. The net effect is that star formation is more likely to occur in some places than it was before the encounter. The starbursts can be observed in the form of radio continuum emission at 20 cm as a result of radiation from supernova remnants, as well as infrared emission. Morphological changes in galaxies as a result of gravitational interactions are described in Chapter 12.

Galaxies that are undergoing collisions or close encounters typically have a change in the star formation rate compared with normal unperturbed galaxies. This effect was first noticed on the basis of galaxy colors; interacting galaxies in the Arp atlas have a wider range of colors around the average for a given Hubble type (nearly ±0.2 mag) than noninteracting galaxies (±0.1 mag). This result is interpreted to mean that some interacting galaxies have excess star formation (colors bluer than normal), while others have a deficit (colors redder than normal). Interactions with distant galaxy cluster members can lead to overall gas deficiencies on a long time scale, which may result from gas stripping due to the interactions. Virgo spirals are presently H I-deficient by as much as a factor of 10 compared with field spirals, with decreased Hα and FIR emission from less star formation. Furthermore, Hα emission is lower in galaxies near the cluster center than in the same Hubble types in the outer parts of the cluster or in the field, signaling less star formation at present.

The star formation efficiency inferred from the ratio of far-infrared luminosity to CO luminosity, L_{FIR}/L_{CO}, is 7 to 8 times higher in interacting and merging galaxies than in isolated galaxies. The star formation efficiency (SFE) in interacting galaxies is highest for galaxies with tidal tails. Figure 11.6 shows the interacting galaxy pair NGC 4485/90; the small galaxy NGC 4490 is a starburst galaxy.

Observations of very massive atomic clouds and regions of star formation in interacting galaxies have been modeled with simulations, for example in the interacting pair IC 2163/N2207 (shown in Figure 7.6), in the "Superantennae" galaxy pair, IRAS 19254–7245 (shown in Figure 11.7), and in the Antennae, NGC 4038/9. Interactions can allow clouds with masses $>10^8$ M_o to form. The star formation often occurs at the tips of tidal arms, sometimes with clumps along the tidal arms as well.

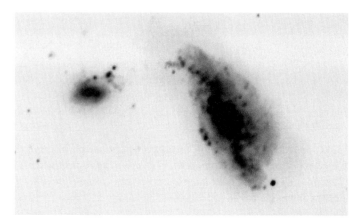

FIGURE 11.6 *NGC 4485/90. (Imaged in B band with the Burrell-Schmidt telescope at KPNO; by the author.)*

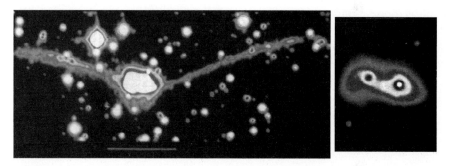

FIGURE 11.7 *The "Superantennae." (left) An infrared image of the interacting galaxies, with clumps visible along the tidal arms. The largest clump is at the end of the right-hand tidal tail. (right) A closeup of the nuclei of the two merging galaxies. (From Mirabel et al. 1991.)*

11.4 Starbursts and the Milky Way

The initial mass function (IMF) in our Galaxy is not perfectly smooth. This observation suggests either that the IMF has varied with time and star formation has proceeded at some constant rate or that the IMF is constant but bursts of star formation occur. In the absence of strong evidence for a variable IMF, the bursting model is a plausible assumption. Figure 11.8 shows the IMF as a function of mass for a constant star formation rate. Now imagine that there is a burst of star formation with a duration τ_B, at a time τ earlier than the present. Then the present-day mass function, PDMF, is not smooth but has a ledge at a mass m_τ whose lifetime is τ, the time since the burst.

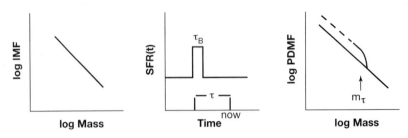

FIGURE 11.8 *Schematic diagram of a smooth initial mass function (IMF) (left), with a burst of star formation (center) combining to produce an IMF with a bump at the mass of the star whose age corresponds to the time since the burst (right). (Adapted from Scalo 1987.)*

Observed ledges in the Milky Way suggest that it has undergone two bursts of star formation, 5×10^9 yr ago (which produced a 1.2 M_o ledge) and 3×10^8 yr ago (which produced a 3 M_o ledge), as shown in Figure 11.9. The Large Magellanic Cloud also appears to have had two bursts of star formation at the same time as those in the Milky Way, but those burst ages are somewhat uncertain. The implication is that two close approaches of the Large Magellanic Cloud and our galaxy may have led to starbursting episodes in the disks. Perhaps the Sun was formed during the older burst.

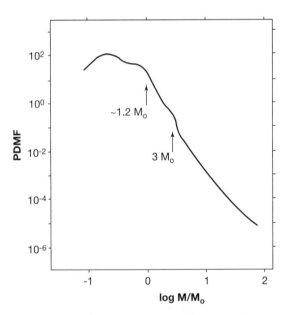

FIGURE 11.9 *Present-day mass function (PDMF) of main sequence stars in the Milky Way, showing possible bursts of star formation twice in the past. (From Scalo 1987.)*

The inference based on statistics of starburst galaxies is that most spirals have undergone a starburst phase. Thus, models that assume a constant star formation rate are probably not accurate over the lifetime of a disk. Population synthesis models generally need to allow for bursting episodes in order to arrive at observed colors in different systems (see Chapter 4).

11.5 Black Hole Activity

Some galaxies have very strong central emission at nonoptical wavelengths, particularly in the ultraviolet or infrared, with additional strong nonthermal radiation. Their nuclear luminosities range from 10^{43} to 10^{48} erg s^{-1}. These galaxies are believed to contain central black holes with accretion disks. They are referred to as *AGNs*, or active galactic nuclei galaxies, and may be any Hubble type. Approximately 1% of otherwise normal spirals have AGNs. Starbursts and AGNs are both very luminous, and it is not easy to sort out whether the excess luminosity is from star formation, accretion disks, or some combination of the two. Many starbursting galaxies may also contain central black holes; indeed, starbursts may provide the fuel for these black holes in the form of massive stellar remnants and infalling supernova debris. An accretion rate of ~1–10 M_0 yr^{-1} is required to provide the energy for the observed x-ray and uv radiation of 10^{12}–10^{13} L_0. In this section we will explore the general properties of black holes, before returning to a discussion of the galaxies in which they are likely to reside.

A black hole is a region inside which gravitational forces are so strong that nothing can escape, even photons. A nonrelativistic approximation for the limiting radius of a black hole is given in terms of its mass by equating gravitational potential energy with kinetic energy and replacing the velocity with the speed of light:

$$\frac{GMm}{R} = \frac{1}{2}mv^2 = \frac{1}{2}mc^2$$

The edge of the black hole, known as the *event horizon*, can be solved in terms of the mass from this equation. Its radius, R_{Sch}, is called the *Schwarzschild radius* in honor of its discoverer and is given exactly by:

$$\boxed{R_{Sch} = \frac{2GM}{c^2}}$$

There is a convenient conversion factor between radius and mass: The radius of a black hole, measured in kilometers, is approximately equal to 3 times the mass of the black hole, measured in solar masses. That is,

$$R_{Sch}(km) = 3M(M_o)$$

For example, a 10 M_o black hole has a radius of 30 km, and a 10^6 M_o black hole has a radius of 3×10^6 km.

There is no limiting density that defines a black hole. We can determine the density of a black hole by recalling that the density is the mass per unit volume, $\rho = M/(4\pi r^3/3)$, and substituting so that the density is expressed in terms of either the mass or the radius. Using the Schwarzschild equation, we have

$$\rho = \frac{M}{\frac{4}{3}\pi \left(\frac{2GM}{c^2}\right)^3} = \frac{3c^6}{32\pi G^3 M^2}$$

Thus, the density of a black hole scales inversely with the square of the mass (or radius); larger black holes are less dense than smaller ones. In fact, a 1 M_o black hole has a density $\sim 10^{18}$ g cm^{-3}, whereas a 10^9 M_o black hole has a density of 1 g cm^{-3}, the same as water!

Although black holes do not radiate, the accretion disks surrounding them do. The rapid motions necessary for gas to orbit lead to thermal temperatures of 10^6 K, corresponding to thermal x-ray emission. Accretion disk emission increases rapidly in intensity in the vicinity of a black hole. There is a limit to the amount of radiation that is generated by matter falling into the accretion disk. This limit, known as the *Eddington limit*, is given by an outward force of radiation scattered off of free elections in the accreting gas. The corresponding Eddington luminosity is given by:

$$\frac{GmM}{R^2} = \frac{\sigma L}{4\pi c R^2}$$

for mass M and radius R of the black hole, mass m of a proton, luminosity L, and electron-photon cross section $\sigma = 7 \times 10^{-25}$ cm^{-2}. Then the luminosity is

$$L_{Edd} \approx 1.3 \times 10^{38} \left(\frac{M}{M_o}\right) \text{ erg s}^{-1}$$

Quasars are observed to radiate at approximately their Eddington limits, supporting the notion that the radiation is from an accretion disk around a black hole; an example is shown in Figure 11.10.

There are strong magnetic fields associated with the accretion disk around a black hole. The geometry is such that a jet is formed perpendicular to the plane

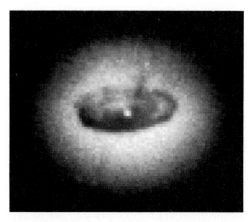

FIGURE 11.10 *NGC 4261 black hole accretion disk. On a larger scale, it has radio jets perpendicular to the disk. (Imaged with Hubble Space Telescope WFPC2 by L. Ferrarese, Johns Hopkins University; and NASA.)*

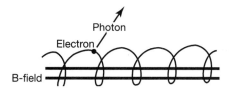

FIGURE 11.11 *Synchrotron radiation resulting from electrons in a magnetic field.*

of the accretion disk, where particles can escape. The magnetic fields lead to strong nonthermal emission in the form of *synchrotron radiation*, which is caused by electrons radiating as they spiral around the magnetic field lines. A schematic diagram is shown in Figure 11.11.

Synchrotron radiation has a power law spectrum, in which the intensity scales as the frequency to some power α and to the magnetic field to a power $\alpha+1$:

$$L_\nu \propto B^{\alpha+1}\nu^{-\alpha}$$

where α is typically 0.4 to 0.8 for energies <5 GeV and ~1 for energies >5 GeV. Synchrotron radiation is distinguishable from blackbody radiation based on the shape of the intensity curve as a function of frequency. Figure 11.12 shows the composite spectrum of a quasar, believed to be the result of blackbody radiation from regions at three different temperatures in the accretion disk, as well as synchrotron radiation following a power law distribution. The suspected central black hole has a mass of $0.2\text{--}0.5 \times 10^9\ M_o$.

The flux density distributions are shown in Figure 11.13 for four active galaxies, also with composite spectra.

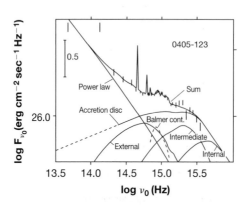

FIGURE 11.12   *The spectrum of quasar 0405–123 can be decomposed into radiation from three different temperature blackbodies, in addition to synchrotron radiation that has a power law. (From Malkan 1983.)*

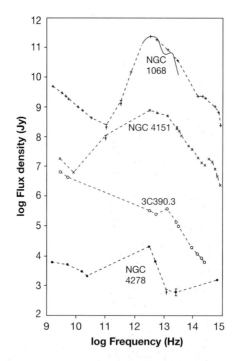

FIGURE 11.13   *Continuum energy distributions for four active galaxies. The straight-line declining parts are from synchrotron radiation; the peaked parts are from blackbody radiation. (From Soifer et al. 1987.)*

The presence of a black hole can be inferred from optical observations of velocities, because a massive central object such as a black hole would cause rapid rotation and high velocity dispersions close to it. Such rapid motion has been ob-

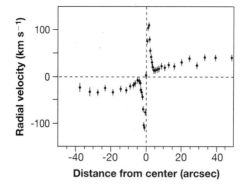

FIGURE 11.14 *M31 central peak, which shows up in the rotation curve. (From Kormendy 1988.)*

served in the core of M31, whose central rotation curve is shown in Figure 11.14, and in several other galaxies.

Hubble Space Telescope observations of the cores of galaxies indicate that black holes may be common. Just as central massive black holes would cause rapid rotation, so they would also lead to an increase in the amount of light coming from the center due to the enhanced stellar densities. Enhanced central

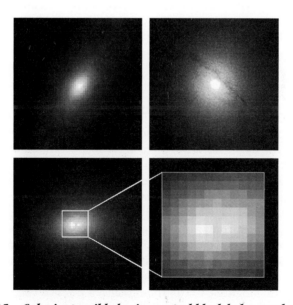

FIGURE 11.15 *Galaxies possibly having central black holes, as observed with Hubble Space Telescope WFPC2: (top left) NGC 3377, 2.7 arcsec box; (top right) NGC 3379, 5.4 arcsec box; (bottom left) NGC 4486B, 2.7 arcsec box; (bottom right) NGC 4486B with double nucleus, 0.5 arcsec box. (Photos from Hubble Space Telescope public dataset; imaged by K. Gebhardt, University of Michigan; and T. Lauer, NOAO.)*

FIGURE 11.16 *Center of M51. (Imaged with Hubble Space Telescope by N. Panagia, STScI and European Space Agency; and NASA.)*

peaks are shown for some nearby galaxies in Figure 11.15 thought to have supermassive black holes; these peaks are not definitive in identifying black holes but are consistent with their presence.

The nearby spiral NGC 5194 has a remarkable pattern of dust crossing into the central regions, with what looks like a black hole accretion disk in the center, as shown in Figure 11.16. The nuclear region is a *LINER* (low ionization nuclear emission-line region) showing outflow.

11.6 The Galactic Center in the Milky Way

The central region of the Milky Way is uniquely problematic to observe because of the high optical opacity; it is observed at radio and infrared wavelengths. The inner 800 pc is mostly molecular, with a total mass of $M_{H_2} \sim 1\text{--}5 \times 10^8 M_\odot$ based on CO observations. Inside 300 pc, the gas is in the form of giant molecular clouds. In what is called the Sgr B2 complex, there is an apparently expanding ring approximately 200–300 pc in radius containing high-density, high-mass clouds. There are seven compact H II regions in this area. Observations of atomic gas reveal a rapidly rotating H I disk with a radius of ~700 pc and a mass of $4 \times 10^6 M_\odot$. At 3 kpc, there is an armlike distribution that is not in circular rotation, and there is an absorption line component moving away from the Galactic Center near a galactic longitude $l = 0°$. There is also a positive velocity gas component moving at 135 km s^{-1} in quadrant IV, where gas motions should show negative Doppler shifts. These observations also suggest an expansion away from the Galactic Center, and the feature is referred to as an *expanding arm*. Some of the unusual velocities may be explained by the galactic fountain model and the small central bar, described in Chapter 8. The rotation axis of both the central H I disk and the "expanding" H I is tilted out of the galactic plane. There is an anomalous velocity "40 km/s cloud" near Sgr A, which is identified with the Galactic Center. There are many streamers and filaments associated with this region, which radiates strong infrared and radio continuum emission. The total stellar mass within 50 pc is $5 \times 10^8 \, M_o$.

Observations at infrared and radio continuum (cm) wavelengths reveal two components near the Galactic Center. One, called *Sgr A East*, is a supernova remnant. The other, *Sgr A West*, is an ultracompact nonthermal source with a size R < 10 AU. Its inferred brightness temperature of T ~10^{10} K suggests nonequilibrium conditions, such as a small black hole accretion disk. A schematic diagram of many of the features is shown in Figure 11.17. The black hole mass is estimated to be 10^3 to $10^6 \, M_o$ based on several different observations. For example, recent lensing observations of light refracted around the Galactic Center suggest a $10^3 \, M_o$ black hole. This mass implies a radius of 3×10^3 km, which would mean that the Galactic Center black hole is only the size of Mars. Recent observations of gamma rays streaming from the Galactic Center perpendicular to the plane indicate positron-electron annihilations, which probably result from activity around the central black hole.

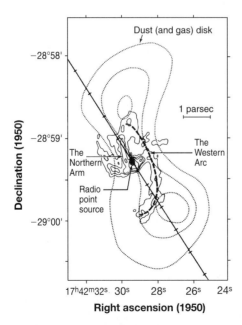

FIGURE 11.17   *The center of the Milky Way, showing minispiral structure in the radio continuum at 2 cm. The point source is Sgr A*. The lobes are 60–100 µ dust radiation. The diagonal line represents the galactic plane. (From Genzel and Townes 1987.)*

11.7 Unified Models of Quasars, Radio Galaxies, Blazars, and Seyferts

Quasars are point sources at high redshift with very intense radiation coming from very small regions; their total luminosities range from $\sim 10^{12}$ to 10^{15} L_o. They were first discovered as stellar-like sources at the locations of strong radio emitters. Because they resembled stars on photographic plates, they were called *quasi-stellar objects*, later abbreviated to QSRs (radio-loud) or QSOs (radio-quiet) and now referred to as quasars. About 10% of quasars are radio-loud. Quasar spectral lines were not identified at first because they were (unexpectedly) highly redshifted; the nearest quasar (3C273) has z = 0.15, but most are at z > 2. Their intensities are consistent with central black holes having masses of 10^9 M_o.

There are currently about 1000 known quasars (see Figure 2.20). A new effort to identify quasars on the basis of their radio emission (the FIRST survey using the Very Large Array radio telescope) has identified several hundred additional candidates. There may be as many as a billion quasars in the visible Universe.

Quasars are sometimes observed to vary optically on time scales of a few days or months. This variability is a clue to their size, since their maximum diameter corresponds to the light travel time across them. A schematic diagram of a light curve for a quasar is shown in Figure 11.18. The oval represents the quasar of diameter d, a distance x away from the observer on the left. Suppose that it starts to vary in brightness. The continuum emission level begins to increase when the light from the near side reaches the observer, at a time $t_1 = x/c$. The brightness returns to the quiescent level when the light from the far side has reached the observer, at a time $t_2 = (x + d)/c$. Then the size of the quasar is $d = (t_2 - t_1)/c$. For example, quasars that vary on a one-day time scale have maximum sizes of 1 light-day, which is a few times the size of the solar system (it takes 6 h for emission from Pluto to reach us). Their corresponding black hole masses are 10^9 M_o.

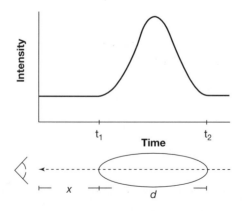

FIGURE 11.18   *A black hole's variability is an indication of its size.*

Emission in other active galaxies such as Seyferts also shows variability, which probably is related to the way in which energy is transferred within an accretion disk. Optical observations of the light curve of the Seyfert galaxy NGC 5548 show similar variations with a slight time lag for the optical light compared with the ultraviolet (Figure 11.19).

Radio galaxies are galaxies with extended symmetric lobes of radiation typically a few hundred kiloparsecs up to 1 Mpc long, as illustrated schematically in Figure 11.20 and shown for Cygnus A in Figure 11.21. The extent of the radio lobes scales with the radio luminosity; brighter sources have larger lobe extents. Radio galaxies are found at lower redshift than quasars, so the host galaxies are easier to observe. Their total luminosities, like those of Seyferts, are typically up to $\sim 10^{12}$ L_o.

The central galaxies usually are ellipticals. An example is Centaurus A, shown in Figure 11.22. This galaxy has an extensive dust lane crossing the central part of the galaxy, probably due to the accretion of a neighboring galaxy. It has jets that are perpendicular to the plane of the dust (which is much larger than the central accretion disk but with the same orientation). Other examples are shown in Figure 2.19; star formation is occurring in the knots of the jet 3C368.

Superluminal jets are sometimes observed in radio galaxies; as the name implies, these are jets that appear to be moving faster than the velocity of light. M87 is an elliptical galaxy with a jet that has been observed at optical and radio wavelengths. An HST high-resolution image of its center is shown in Figure 11.23.

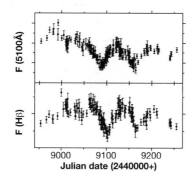

FIGURE 11.19 *Variability in the light curves of NGC 5548 from optical (top) and Hβ (bottom) observations. (Korista et al. 1995).*

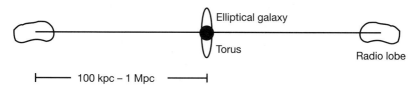

FIGURE 11.20 *Radio lobe galaxies have extended radio emission far from the optical galaxy but centered on it.*

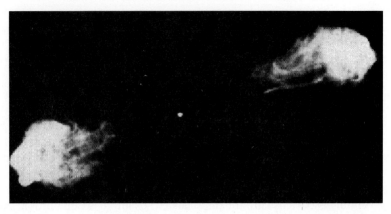

FIGURE 11.21   *Cygnus A with 6-cm radio emission lobes. (From Perley et al. 1984.)*

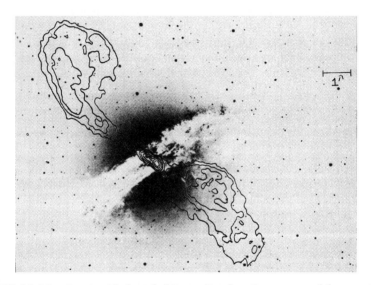

FIGURE 11.22   *Cen A with dust (white regions) and contours of 6-cm emission showing radio lobes. (From Burns et al. 1983.)*

Radio jets are concentrations of gas that produce synchrotron radiation; observed over a period of years, distinct blobs in them shift their relative positions in a systematic outflow from the nucleus, as shown in Figure 11.24 for another source.

If the apparent line along which the blobs move has a very small angle to our line of sight, and their motion is toward us at close to the speed of light, then the apparent motion may exceed the speed of light. Consider Figure 11.25, in which a blob of material travels from point A to point B along a line that has an angle θ from our line of sight. Since point B is closer to us, its emission takes a shorter time to reach us than the emission from point A, so the apparent time it took the

FIGURE 11.23 *(left) Nucleus of M87. (right) Jet of high-speed electrons. (Imaged with the Hubble Space Telescope WFPC2 by H. Ford, STScI/Johns Hopkins University; R. Harms, Applied Research Corp.; Z. Tsevetanov, A. Davidsen, and G. Kriss: Johns Hopkins University; R. Boblin and G. Harlig, STScI; L. Dressel and A. Kochbar, Applied Research Corp.; B. Margon, University of Washington; and NASA.)*

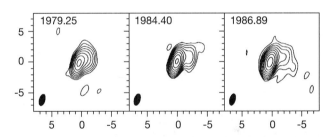

FIGURE 11.24 *Shift of blobs with time in quasar 1633+382. (From Barthel et al. 1995.)*

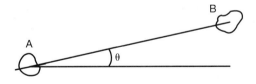

FIGURE 11.25 *Superluminal jet geometry; θ is the angle between the jet motion and our line of sight.*

blob to go from A to B is shorter than the time it really took. The ratio of the apparent velocity to the true velocity is given by

$$\frac{v_{apparent}}{v_{true}} = \frac{1}{\sqrt{1 - (\cos \theta)^2}}$$

where θ is the angle between the path of the blob's motion and our line of sight. If the angle is small and the true velocity close to the speed of light, then the apparent velocity exceeds the speed of light.

X-ray observations indicate that there is hot gas between galaxies in clusters, known as the *intracluster medium* (ICM). It is possible that interactions between the ICM and a galaxy's ISM (interstellar medium) trigger massive star formation. This triggering is most likely to occur in galaxies encountering a dense part of the ICM for the first time, before significant amounts of gas are stripped from the galaxy. The physical mechanism for ISM-ICM triggering is thought to be the ram pressure causing a compression of clouds in the galaxy. If the ICM has a density less than 10^{-4} cm^{-3}, then magnetic and turbulent forces supporting gravitationally bound molecular clouds exceed the ICM-ISM ram pressure. In this case, ICM-ISM interactions could compress low-density H I clouds into molecular clouds. If the ICM density exceeds $\sim 10^{-3}$ cm^{-3}, the ram pressure could compress pre-existing molecular clouds into star formation.

Cooling flows in clusters of galaxies and in elliptical galaxies involve hot gas that has been ejected by young stars and then cooled by x-ray emission from thermal *Bremsstrahlung radiation*, also known as *free-free radiation*. Here, electrons accelerate and radiate in the presence of ions. The gas emissivity $j(\nu)$ is given approximately by

$$4\pi j(\nu) = 6.79 \times 10^{-38}\, n_e^2\, T^{-1/2}\, e^{-h\nu/kT}\, \text{erg cm}^{-3}\, \text{sec}^{-1}\, \text{Hz}^{-1}$$

for electron density n_e and gas temperature T. Typical temperatures of the intragalactic gas are 10^8 K, with densities of 10^{-3} cm^{-3}. The thermal velocity of the gas is $\sim 10^3$ km s^{-1}, which is approximately the same as the galaxy motions in the cluster. The gas is typically condensing at the rate of about 1 M_o yr^{-1} kpc^{-1} as it cools. Theoretical models indicate that a few percent of the cooled gas may form low-mass stars.

Figure 11.26 shows a jet from the elliptical galaxy NGC 1265 interacting with the intragalactic gas; the galaxy is moving at 2500 km s^{-1} through the cluster, so the jet is swept back. The morphology of the jet structure is sometimes called a *head-tail source*. The spectral index α of the emission is approximately -0.65 between 2 and 21 cm, where the intensity $S \propto \nu^\alpha$; the spectral index changes slightly across the knots in the jet. The jet structure is polarized, with the magnetic field aligned along the jet in the inner tail and perpendicular to it in parts of the outer tail.

Seyfert galaxies, as mentioned in Chapter 2, are spiral galaxies that appear normal except for their intense nuclear emission, which accounts for approximately half of their total optical luminosity. Most Seyferts are too far away to sort out the relative contributions of young stars and black holes to the infrared emission, although careful studies of emission lines and continuum spectra yield some clues. Their nuclear spectra show high excitation emission lines superposed on a red continuum, in contrast to the blue continuum of starbursts. Seyfert type 1

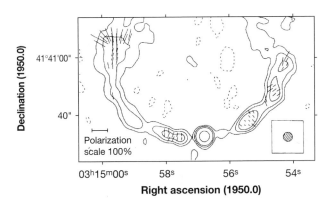

FIGURE 11.26 *Gas flow around NGC 1265, showing the jet tails; short lines represent the electric field orientation, which is perpendicular to the magnetic field. (From O'Dea and Owen 1986.)*

galaxies and type 2 galaxies differ primarily in having broad wings or no wings, respectively, in their hydrogen emission lines. The broad wings evidently originate from fast-moving discrete clouds in what is called the *broad-line region* (*BLR*), which is just outside the accretion disk. Narrower lines come from an extended region (*narrow-line region*, or *NLR*), as much as 1 kpc in diameter. The NLR has densities of $\sim 10^3$ cm^{-3}, whereas the BLR has densities in excess of 10^9 cm^{-3} and temperatures of a few times 10^4 K. Shock ionizations between colliding clouds may account for some emission variability. When there is variability in the continuum and line emission, there is often a time lag between the two, but the mechanism is not well understood at this time. The presence or absence of wings may be associated with the opacity of an accretion disk and the angle at which we are viewing the nuclear region; Seyfert 2s evidently have a thicker, denser torus than Seyfert 1s.

NGC 1068 is the nearest galaxy with a Seyfert 2 active nucleus and also has a high star formation rate in the inner disk (see Figure 2.18 and Figure 11.27). The

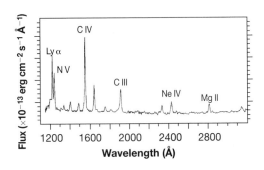

FIGURE 11.27 *Spectrum of NGC 1068, a Seyfert 2 galaxy. (From Kinney et al. 1993.)*

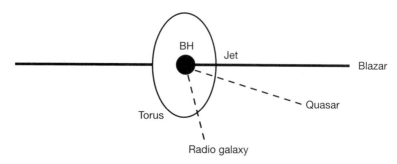

FIGURE 11.28 *Unified model, in which active galaxies are all the same type of object (i.e., a galaxy with a central black hole and an accretion disk) viewed from different aspects.*

total luminosity of $L_{tot} = 2 \times 10^{11}$ L_o within 1.5 kpc of the galactic center is from young stars, and from a compact source centered on the Seyfert nucleus. CO observations show two inner spiral arms going to a ring at 1.5 kpc, with a central bar. There are several distinct gas cloud complexes 500 pc in size, with masses totaling 7×10^8 M_o. Within the innermost 130 pc, the total amount of molecular hydrogen gas is ~8×10^7 M_o.

BL Lacertae objects, sometimes called *BL Lac objects* or *blazars*, have quasar-like activity. They have no emission lines but are highly variable. As mentioned in Chapter 2, they are active elliptical galaxies. Thus, they are the analogs of Seyferts, which are active spiral galaxies.

Observations of Seyferts, quasars, radio galaxies, and blazars all suggest the presence of black holes in the central regions. The similarity between the different objects lends itself to interpretation in terms of a *unified model*, in which all are galaxies with central accretion disks and black holes viewed from different aspects. Figure 11.28 illustrates the way in which an object with a central black hole and torus would appear from different lines of sight. If we can see into the core region, a very bright blazar appears if the galaxy is elliptical (or a Seyfert 1 if it is spiral). A slight viewing angle would put us off-axis of the primary emission, so the object might appear as a quasar. A larger viewing angle would allow us to see radio lobes from outflow.

In Chapter 12, we will examine the large-scale structure of galaxies in the Universe and explore their formation.

Exercises

1. Determine the Eddington limit for a black hole with a radius of 10^{13} cm.
2. If a superluminal jet has a peak that moves 5 milliarcsec with a maximum apparent velocity of 3.5c, what is the jet's true velocity?
3. A quasar at z = 0.3 is observed to have emission at 1400 GHz with an intensity of 10^{28} erg cm^{-2} sec^{-1} Hz^{-1}, which you can assume is due to synchrotron emission. It has

thermal emission from an accretion disk with a temperature of 10^6 K. Use a spreadsheet or write a program to plot the resultant combined spectrum.

4. Determine the Schwarzschild radius of a black hole whose density is 10^{-6} g cm^{-3}.

5. If there is a PDMF peak at a mass of 5 M$_0$ as a result of an interaction between two galaxies, when did the closest approach occur?

Unsolved Problems

1. How can we discriminate between central black holes and star formation?
2. What are the details of mass transfer that supply power to AGNs?

Useful Websites

The latest HST images of black holes are on http://www.stsci.edu

Kennicutt's spectrophotometric atlas of galaxies is available through the Astronomical Data Center: http://adc.gsfc.nasa.gov

Further Reading

Antonucci, R. 1993. Unified models for active galactic nuclei and quasars. *Annual Review of Astronomy and Astrophysics* 23:147.

Athanassoula, E., and A. Bosma. 1985. Shells and rings around galaxies. *Annual Review of Astronomy and Astrophysics* 23:147.

Barnes, J., and L. Hernquist. 1992. Dynamics of interacting galaxies. *Annual Review of Astronomy and Astrophysics* 30:705.

Barthel, P., J. Conway, S. Myers, T. Pearson, and A. Readhead. 1995. New superluminal quasar 1633+382 and the blazar–gamma-ray connection. *Astrophysical Journal Letters* 444:L21.

Benedict, G., B. Smith, and J. Kenney. 1996. CO in NGC 4314: The detection of inflow along the bar. In *IAU Colloquium 157: Barred galaxies, ASP Conference Series, 91*, ed. R. Buta, B. Elmegreen, and D. Crocker, 227. San Francisco: Astronomical Society of the Pacific.

Blandford, R., H. Netzer, L. Woltjer, T. Courvoisier, and M. Mayor. 1990. *Active Galactic Nuclei.* New York: Springer-Verlag.

Bowers, R., and T. Deeming. 1984. *Astrophysics II: Interstellar matter and galaxies.* Boston: Jones and Bartlett Publishers.

Bridle, A., and R. Perley. 1984. Extragalactic radio jets. *Annual Review of Astronomy and Astrophysics* 22:319.

Burns, J., E. Feigelson, and E. Schreier. 1983. The inner radio structure of Centaurus A: Clues to the origin of the jet x-ray emission. *Astrophysical Journal* 273:128.

Elmegreen, B. 1994. Theory of starburst and ultraluminous galaxies. In *Violent star formation from 30 Dor to QSOs*, ed. G. Tenorio-Tagle, 220. Cambridge: Cambridge University Press.

Genzel, R., and C. Townes. 1987. Physical conditions, dynamics, and mass distribution in the center of the galaxy. *Annual Review of Astronomy and Astrophysics* 25:377.

Giavalisco, M., C. Steidel, and F. Macchetto. 1996. Hubble Space Telescope imaging of star-forming galaxies at redshifts z>3. *Astrophysical Journal* 470:189.

Gregg, M., R. Becker, R. White, D. Helfand, R. McMahon, and I. Hook. 1996. The FIRST bright QSO survey. *Astronomical Journal* 112:407.

Ho, L., A. Filippenko, and W. Sargent. 1995. A search for dwarf Seyfert nuclei. 2. An optical spectral atlas of the nuclei of nearby galaxies. *Astrophysical Journal (Suppl.)* 98:477.

Holtzman, J., et al. 1996. Star clusters in interacting and cooling flow galaxies. *Astronomical Journal* 112:416.

Kenney, J., 1996. Molecular gas in the central regions of barred galaxies. In *IAU Colloquium 157: Barred galaxies, ASP Conference Series, 91*, ed. R. Buta, B. Elmegreen, and D. Crocker, 150. San Francisco: Astronomical Society of the Pacific.

Kennicutt, R. 1992. A spectrophotometric atlas of galaxies. *Astrophysical Journal (Suppl.)* 79:255.

Kinney, A., R. Bohlin, D. Calzetti, N. Panagia, and R. Wyse. 1993. An atlas of ultraviolet spectra of star-forming galaxies. *Astrophysical Journal (Suppl.)* 86:5.

Korista, K. 1992. The broad emission-line profiles and profile variability of the Seyfert 1 galaxy Arakelian 120. *Astrophysical Journal (Suppl.)* 79:285.

Korista, K., et al. 1995. Steps toward determination of the size and structure of the broad-line region in active galatic nuclei. 8. An intensive HST, IUE, and ground-based study of NGC 5548. *Astrophysical Journal (Suppl.)* 97:285.

Kormendy, J. 1988. Evidence for a supermassive black hole in the nucleus of M31. *Astrophysical Journal* 325:128.

Kormendy, J., and D. Richstone. 1995. Inward bound—The search for supermassive black holes in galactic nuclei. *Annual Review of Astronomy and Astrophysics* 33:581.

Kunth, D., T.X. Thuan, and J. Tran Thanh Van. 1985. *Star-forming dwarf galaxies and related objects.* Paris, France: Editions Frontieres.

Lauer, T., S. Tremaine, E. Ajhar, R. Bender, A. Dressler, S. Faber, K. Gebhardt, C. Grillmair, J. Kormendy, and D. Richstone. 1996. Hubble Space Telescope observations of the double nucleus of NGC 4486B. *Astrophysical Journal Letters* 471:L79.

Lauer, T., E. Aghar, Y.-I. Byun, A. Dressler, S. Faber, C. Grillmair, J. Kormendy, D. Richstone, and S. Tremaine. 1995. The centers of early-type galaxies with HST. I. An observational survey. *Astronomical Journal* 110:2622.

Longair, M. 1997. Active galaxies: The redshift one 3CR galaxies. *Astronomy and Geophysics* 38:10.

Malkan, M. 1983. The ultraviolet excess of luminous quasars. II. Evidence for massive accretion disks. *Astrophysical Journal* 268:582.

Menon, T. 1995. Starburst phenomena associated with tidal action in a spiral galaxy in a compact group. *Astronomical Journal* 110:2605.

Meurer, G., T. Heckman, C. Leitherer, A. Kinney, C. Robert, and D. Garnett. 1995. Starbursts and star clusters in the ultraviolet. *Astronomical Journal* 110:2665.

Mirabel, I., D. Lutz, and J. Maza. 1991. The "Superantennae." *Astronomy and Astrophysics* 243:367.

Moffet, A.T. 1975. Strong nonthermal radio emission from galaxies. In *Galaxies and the universe*, ed. A. Sandage, 212. Chicago: University of Chicago Press.

Norman, C., and N. Scoville. 1988. The evolution of starburst galaxies to active galactic nuclei. *Astrophysical Journal* 332:124.

O'Dea, C., and F. Owen. 1986. Astrophysical implications of the multifrequency VLA observations of NGC 1265. *Astrophysical Journal* 316:950.

Osterbrock, D. 1993. The nature and structure of active galactic nuclei. *Astrophysical Journal* 404:5510.

Osterbrock, D., and W. Mathews. 1986. Emission-line regions of active galaxies and QSOs. *Annual Review of Astronomy and Astrophysics* 24:171.

Perley, R., J. Dreher, and J. Cowan. 1984. The jet and filaments in Cygnus A. *Astrophysical Journal Letters* 285:L35.

Peterson, B. 1993. Reverberation mapping of active galactic nuclei. *Publications of the Astronomical Society of the Pacific* 685:247.

Rees, M. 1984. Black hole models for active galactic nuclei. *Annual Review of Astronomy and Astrophysics* 22:471.

Salzer, J., J. Moody, J. Rosenberg, S. Gregory, and M. Newberry. 1995. Imaging and spectroscopic observations of the case survey blue/emission-line galaxies. *Astronomical Journal* 109:2376.

Sanders, D., E. Phinney, G. Neugebauer, B. Soifer, and K. Mathews. 1989. Continuum energy distributions of quasars—Shapes and origins. *Astrophysical Journal* 347:29.

Scalo, J. 1987. The initial mass function, starbursts, and the Milky Way. In *Starbursts and galaxy evolution*, ed. T. Thuan, T. Montmerle, and J. Tran Thanh Van. 445. Paris, France: Editions Frontieres.

Soifer, B., J. Houck, and G. Neugebauer. 1987. The IRAS view of the extragalactic sky. *Annual Review of Astronomy and Astrophysics* 25:187.

Telesco, C. 1988. Enhanced star formation and infrared emission in the centers of galaxies. *Annual Review of Astronomy and Astrophysics* 26:343.

Thronson, H., and J.M. Shull. 1990. *Second Wyoming Conference: The interstellar medium in galaxies*. Dordrecht: Kluwer Academic.

Thuan, T., C. Balkowski, and J. Tran Thanh Van. 1992. *Physics of nearby galaxies: Nature or nurture?: XIIth Moriond Astrophysics Meeting.* Paris, France: Editions Frontieres.

Thuan, T., T. Montmerle, and J. Tran Thanh Van. eds. 1987. *Starbursts and galaxy evolution*. Paris, France: Editions Frontieres.

Tomisaka, K., and J. Bregman. 1993. Hot gas outflows from starburst galaxies. In *Star formation, galaxies, and the interstellar medium*, ed. J. Franco, F. Ferrini, and G. Tenorio-Tagle. 265. Cambridge: Cambridge University Press.

Unavane, M., R. Wyse, and G. Gilmore. 1996. The merging history of the Milky Way. *Monthly Notices of the Royal Astronomical Society* 278:727.

CHAPTER 12
Large-Scale Distributions

Chapter Objectives: to explore the spatial distributions of different types of galaxies as a function of environment

---

Toolbox:

luminosity function   correlation parameter
standard candles

12.1 Galaxy Luminosity Function

Galaxies tend to occur in groups and clusters, which makes them relatively close to each other compared with their diameters—analogous to pennies separated by meters or less. We can examine their distributions by considering a *galaxy luminosity function* $\phi(L)$, which is the number of galaxies for a given luminosity range dL in a given volume. The luminosity function has been computed for many different clusters, with the result that the number of galaxies declines steadily with increasing luminosity. This is the same type of relationship as found for stars.

The luminosity function generally is written in a form known as the *Schechter luminosity function*, $\phi(L)$, derived from a theoretical analysis of gravitational condensations:

$$\Phi(L)dL = n^*\left(\frac{L}{L^*}\right)^\alpha \exp\left(\frac{-L}{L^*}\right)\frac{dL}{L^*}$$

The function matches cluster observations well and is approximately constant but varies slightly from cluster to cluster. The power α is the slope of the low luminosity end; the Schechter fit was $\alpha = -1.25$, but recent fits to more distributions indicate a range from -1.25 to -1.5. In field galaxies, the power law fit is closer to -1, which may be explained by a higher fraction of dwarf elliptical galaxies in denser groups. L^* is the *characteristic luminosity*, where the slope of the function changes rapidly; it is $\approx 1.0 \times 10^{10}$ h^{-2} L$_o$. For a Hubble constant H$_o$ = 50 km s^{-1} Mpc^{-1}, L$^* \approx 3.4 \times 10^{10}$ L$_o$ in V band, which corresponds to the *characteristic magnitude* M$_v^* = -21.5$, but it ranges from ~ -23 to -20 in different clusters. The normalization is n$^* \approx 1.2 \times 10^{-2}$ h^3 Mpc^{-3} for h = 0.5 to 1.0, which is the ratio of the Hubble constant to 100 km s^{-1} Mpc^{-1}. The observed luminosity function for 20 Abell clusters is shown in Figure 12.1.

The majority of galaxies have a luminosity L < 3L*; in contrast, cDs have 10L*. The Schechter luminosity function is probably an accurate representation for most ellipticals and spirals. On the other hand, it is not certain whether irregulars fit a low luminosity extension of the normal galaxy function or whether the slope decreases or increases at the faint end. The low luminosity end is extremely important to understand, because its slope affects the total mass in the Universe and therefore has cosmological implications. Based on observations of Virgo cluster galaxies and of the Hercules and Perseus-Pisces superclusters (see Section 12.8), the luminosity function may vary both with morphological type and with local density.

In the 1970s, H. Butcher and A. Oemler photographed distant clusters and noticed a preponderance of blue galaxies compared with nearby clusters. This result is called the *Butcher-Oemler effect*. Until the Hubble Space Telescope high-resolution images were available, it was uncertain whether these blue galaxies were different from local galaxies. Now it appears that they are spiral galaxies, but they are unusually blue because of increased star formation in the past. Many of these distant galaxies have unusual asymmetries and rings, perhaps signifying the effects of interactions. Figure 12.2 shows an HST deep image with many faint blue galaxies undergoing intense star formation. Some may be S0 galaxies at a

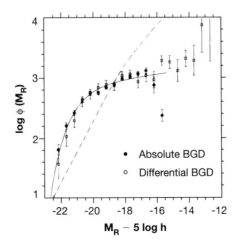

FIGURE 12.1 *Luminosity function for 20 Abell clusters. The dashed line represents the background, BGD. (From Gaidos 1997.)*

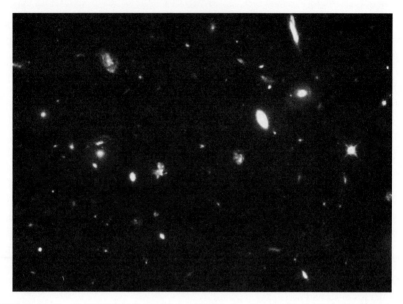

FIGURE 12.2   *Faint blue irregular galaxies in the medium-deep field survey by Hubble Space Telescope WFPC2, based on a 1-day exposure. (Image by R. Windhorst and S. Driver, Arizona State University; W. Keel, University of Alabama; and NASA.)*

more active star formation stage. Theories of galaxy interaction suggest that S0 galaxies were originally spiral galaxies but were stripped of their gas while traveling through a cluster, due to gravitational effects. We will discuss further consequences of interactions in Section 12.6.

In a magnitude-limited sample—that is, where all galaxies are included down to a certain limiting apparent magnitude—a bias is introduced into the data that makes statistical interpretations difficult. Low luminosity galaxies are only sampled nearby, whereas higher luminosity galaxies can be observed to larger distances. Thus, magnitude-limited samples have a disproportionate share of intrinsically brighter members. This effect is known as the *Malmquist bias*. Careful corrections must be made to such samples. A volume-limited sample, where all galaxies are considered within a given distance, is one way of avoiding the Malmquist bias.

12.2 Secondary Distance Indicators

In order to understand the structure of the Universe on a large scale, it is crucial to have accurate distance measurements. In Chapter 3, we saw that parallax is a direct trigonometric measure of distance but can only be applied to objects within about 100 pc of the Sun. Spectroscopic parallax is useful for bright stars and can be applied to stars in nearby galaxies to about 3 Mpc. To extend the distance scale to distant galaxies requires more indirect methods. There are a number of such *secondary distance indicators*, also called *standard candles*, which are categories of objects whose absolute brightnesses are relatively well known or inferred. Their absolute magnitudes are subtracted from the apparent magnitudes to determine the distance modulus. Among the objects useful for this purpose are Cepheids, whose regular brightness variation has a periodicity that depends on the intrinsic brightness (see Chapter 3), as shown in Figure 12.3 for stars in our Galaxy and in M81. This period-luminosity law depends on radial oscillations of the star and the period $P \propto L^{4/3}$ for luminosity L. According to theory, the relation between the absolute bolometric magnitude M_{bol} and the period P is given by

$$M_{bol} = -3.125 \log P - 1.525$$

There is also a color term, which is small because the temperature range of stars in the instability strip is small. Metallicity dependences may also change the coefficients slightly. The narrowness of the period-luminosity-color relation is analogous to the fundamental plane of elliptical galaxies.

There are two types of Cepheids, Type I (also called *classical Cepheids*) and Type II (also called *W Virginis stars*), which can be distinguished by their periods and the shapes of their light curves. Type I Cepheids, which are more massive, have periods of about one day and light curves that are symmetric. Type II Cepheids have periods of about 10 days and light curves that are asymmetric. Type I Cepheids are brighter than Type II by 1.5 mag. Their absolute magnitudes range from $M_V = -2$ to -7, so they provide distance measurements out to nearly 20 Mpc with HST observations.

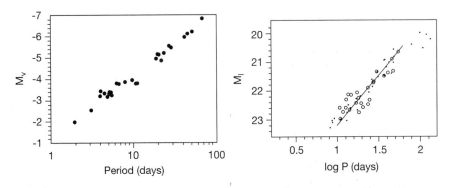

FIGURE 12.3   *(left) P-L relation for Cepheids in the Milky Way. (Data from Feast and Walker 1987.) (right) P-L relation for Cepheids in M81 (open circles) and LMC (dots). (Freedman et al. 1994.)*

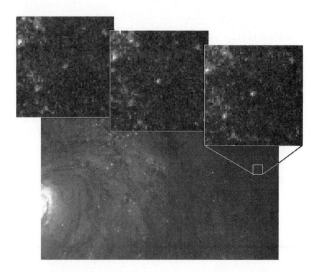

FIGURE 12.4   *Cepheids in M100. The top three images are snapshots of the same region, taken several days apart. The Cepheid is in the center of each frame and varies in brightness. (Photographed with the Hubble Space Telescope WFPC2 by W. Freedman, Observatories of the Carnegie Institution of Washington; and NASA. Images from the HST public dataset.)*

Figure 12.4 shows Cepheids in M100 (NGC 4321). Cepheids are extremely important for calibrating more-distant indicators.

Sandage and R. Humphreys studied M giants in M33 and found that their maximum absolute brightness is about -8 mag. Thus, these red stars are another useful distance indicator out to at least 30 Mpc. As we depend more and more on statistical arguments, the measurement of distance becomes more uncertain. For example, in large clusters of galaxies, the largest galaxies have a fairly constant absolute magnitude from cluster to cluster, with variations of a factor of 2; this near-constancy provides a crude measure of a cluster's distance even out to 10^9 pc.

There are several distance indicators that can measure distances out to about 100 Mpc. In Chapter 7, we discussed the Tully-Fisher relation for spiral galaxies and the $D_n - \sigma$ relation for elliptical galaxies, in which internal motions are correlated with intrinsic brightnesses. The absolute brightness of Type I supernovae is about -19, so they can be viewed out to distances of 2×10^9 Mpc. Novae also have peak brightnesses, which makes them good standard candles.

Another distance method is based on the statistics of luminosities of stellar populations and depends on fluctuations in surface brightness; a galaxy that is more distant appears smoother than an identical closer one. The surface brightness, or flux per pixel, is independent of distance, while the variance in flux scales as the inverse square of the distance and the rms scales with inverse distance. By measuring the ratio of the mean flux $N\bar{f}$ to its *variance* in a region, $N\bar{f}^2$, for number of stars N, the fluctuation flux $\bar{f}$ can be determined and so the

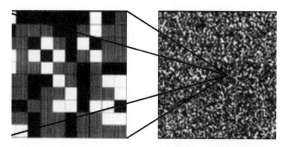

FIGURE 12.5 *Pixel-to-pixel fluctuations are larger for nearby galaxies than for more-distant galaxies. (Image from the Massachusetts Institute of Technology website [Tonry].)*

distance can be inferred. This method is useful for elliptical and S0 galaxies, especially in I band, which minimizes dust absorption. Figure 12.5 illustrates pixel-to-pixel fluctuations at high and low resolution, simulating near and distant galaxies.

Globular clusters provide another method for measuring distances. Absolute visual magnitudes of individual clusters range from an average of $M_V = -7$ up to $M_V = -11$. Elliptical galaxies contain hundreds or even thousands of globular clusters. The *globular cluster luminosity function* (GCLF), or relative number of clusters as a function of magnitude, has the same functional form in essentially all galaxies:

$$\Phi(m) = Ae^{-(m-m_0)^2/2\sigma^2}$$

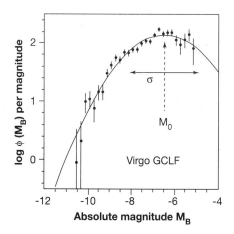

FIGURE 12.6 *Globular cluster luminosity function for four Virgo ellipticals. The vertical scale is arbitrary. (From Harris et al. 1991.)*

where m_0 is the peak magnitude, or turnover point, of the population of clusters, A is the total population of clusters in a galaxy, and σ is the standard deviation of the curve, as shown in Figure 12.6. Empirically, the absolute magnitude of the turnover point for clusters is essentially the same for different galaxies, $M_0 = -6.6 \pm 0.26$. For the Milky Way, the turnover is at $M_0 = -6.8 \pm 0.17$.

A similar luminosity function can also be established for planetary nebulae for their use as extragalactic distance indicators. Planetary nebulae may be identified through their presence in images using narrow-band filters centered on their prominent oxygen emission at $\lambda = 5007$ Å.

12.3 The Hubble Constant

On the basis of HST observations of many different distance indicators whose distance ranges overlap, observers are converging on a value of the Hubble constant below ~75 km s^{-1} Mpc^{-1} (one team estimates 68–78, while another estimates 57). There is considerable debate over the exact value, which is of fundamental importance in understanding the evolution and fate of our Universe. Figure 12.7 shows a Hubble diagram for distant objects. Note that the slope of such a line is the Hubble constant.

The Universe is slowing down in its expansion, because it contains matter whose gravity counteracts the Big Bang explosion. The rate of deceleration depends on the amount of mass in the Universe, so a measure of the deceleration is an indication of whether the Universe will expand forever.

The *deceleration parameter* q_0 is given by

$$q_0 = \frac{- R_0 \ddot{R}_0}{\dot{R}_0^2}$$

where R_0 is the current size of the Universe and $\dot{R}_0$ and $\ddot{R}_0$ are its first and second derivatives. It can also be related to the Hubble constant and the current density ρ_0 by

$$q_0 = \frac{4\pi G \rho_0}{3 H_0^2}$$

The Universe is said to be *spherical* and will collapse if $q_0 > 1/2$, while it is *hyperbolic* and will expand forever if $q_0 < 1/2$; $q_0 = 1/2$ corresponds to a flat universe which will barely expand forever, reaching zero velocity at infinite time. A value of -1 corresponds to a *steady state* cosmology, in which the density of the Universe does not change. Confirmation of the blackbody nature of the cos-

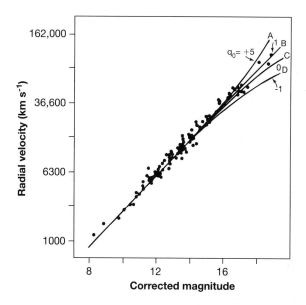

FIGURE 12.7   *Observations of the expansion of the Universe, as measured from motions of individual galaxies, are indicated by dots. Curves show different values of the deceleration parameter. (From Contopoulos and Kotsakis 1987.)*

mic microwave background has eliminated the steady state model as a reasonable possibility. Different curves for q_o are shown in Figure 12.7.

Observations based on quasars indicate a value of q_o = 1.6. Because quasar distances are uncertain, this measurement is not a firm indication of the fate of the Universe; in fact it contradicts density estimates that imply that the Universe is open or nearly flat.

These limits for q_o are equivalent to ratios Ω (the density divided by the critical density; Chapter 7) >1, <1, or = 1, respectively, provided the *cosmological constant* $\Lambda = 0$. The cosmological constant is a hypothetical pressure term that would act to propel the expansion of the Universe (like a negative gravity term); Einstein used it in his initial equations, then retracted it, but recent studies indicate that it may be nonzero.

The age of the Universe can be detemined if the deceleration parameter is known. Note that the Hubble constant has the units of inverse time; thus, for a Universe whose deceleration is 0, the age is given by the inverse of the Hubble constant, $t_H = 1/H_o$. This age is known as the *Hubble time*. For $\Lambda = 0$, the age of the Universe is 2/3 t_H for q_o = 1/2, and less than 2/3 t_H for q_o>1/2. For q_o < 1/2, the age is between 2/3 and 1 t_H. A Hubble constant of 75 km s^{-1} Mpc^{-1} yields a Hubble time of 13 billion years. Whatever the exact value of the Hubble constant, it must be reconciled with the fact that there are old globular clusters whose ages are measured to be 15 to 16 billion years.

12.4 Correlation Function

Galaxies exist primarily in groups, clusters, and rich clusters of galaxies; there are some relatively isolated galaxies, which we call *field galaxies*. There are very few extremely large galaxy clusters, but there are lots of smaller ones. There is some evidence that galaxies formed before clusters.

As a cluster ages, it *relaxes*, meaning that it equilibrates so that the velocities of the individual galaxies have a normal distribution about an average value. The virial theorem shows that the *crossing time* of a cluster (the time it takes for a galaxy to move a distance equal to the cluster diameter) decreases with increasing mass, because then the random velocities of the galaxies are greater. More-massive galaxies tend to concentrate toward the cluster center, and outlying members may be ejected. There is some speculation that our Local Group is a far-flung member of the Virgo cluster.

Separations between pairs of galaxies have been determined on the basis of distributions in the Shane-Wirtanen Catalogue of 1 million galaxies out to nearly 10^3 Mpc, as shown in Figure 12.8, and in more recent observations. Studies in the 1970s were based on galaxies brighter than 17th magnitude; they were not a redshift-selected sample but were just based on the projected distribution of galaxies. Studies of the 1980s and 1990s include a three-dimensional aspect because redshifts have been obtained for large samples of galaxies. The probability that a galaxy has a companion within a given distance is observed to decrease as that distance increases and is specified by a *correlation function*. The separations may be measured in terms of angles or linear dimensions; traditionally, the angular correlation function is called w(θ), while the linear correlation function (the *2-point*

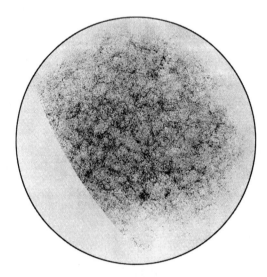

FIGURE 12.8   *The distribution of a million galaxies brighter than 17th magnitude is shown from the Shane-Wirtanen Catalogue (Seldner et al. 1977).*

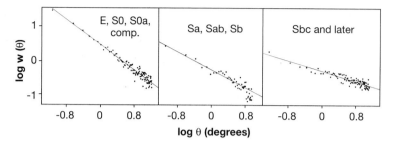

FIGURE 12.9 *Galaxy correlation function log w versus angle for different Hubble types shows tighter clustering for earlier-type galaxies. (Adapted from Haynes and Giovanelli 1988.)*

correlation function or *covariance function*) is $\xi(r)$. The angular correlation function can be converted into a 2-point correlation function if distances are known. For randomly distributed galaxies with a mean density $<n>$, the probability of finding a galaxy in a given volume dV is $<n>$ dV. For inhomogeneously distributed galaxies, the probability that two galaxies will be separated by a distance r within dV is

$$dP = <n> [1 + \xi(r)]dV$$

Based on galaxy counts, the covariance function is $\xi(r) \propto (r/r_o)^{-1.8}$, where r_o is the *correlation length*; it is measured to be ~ 5 h^{-1} Mpc, which indicates that there is significant clustering on scales smaller than this. On the large end, there are also thin filaments and voids on scales of ~ 100 h^{-1} Mpc. There are no preferential scale sizes; this result implies a hierarchy of clustering sizes, as observed in groups, clusters, and superclusters of galaxies (discussed in later sections).

The correlation function depends on Hubble type, as shown in Figure 12.9 for angular separations. The steepness of the slopes is in the sense that E and early-type galaxies are the most tightly clustered, and types Sbc and later are the least tightly clustered. Furthermore, it appears that dwarfs tend to avoid dense clusters but are preferentially in the vicinity of bright galaxies. Recent work indicates that *emission line galaxies*, which may be periodic starbursting dwarfs that are low surface brightness galaxies when they are not starbursting, tend to avoid clusters. There are models of *biased galaxy formation*, in which the luminosity function of galaxies is influenced by the local mass density. In this case, the galaxy luminosity function could differ for clusters and voids. Whether biased galaxy formation actually occurs is as yet unknown.

12.5 Local Group

There are more than two dozen galaxies that are clustered together with the Milky Way; collectively, these are known as the *Local Group*. Their distribution

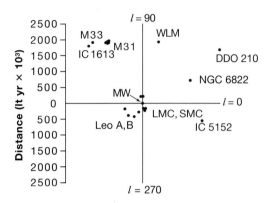

FIGURE 12.10 *Distribution of galaxies in the Local Group, shown on a polar plot with angles giving the galactic longitudes. The Milky Way is at the origin.*

is shown in Figure 12.10. There are 7 galaxies with absolute magnitudes brighter than -16, and at least 21 dwarfs fainter than -16. Andromeda (also known as M31 or NGC 224) and the Milky Way are the only large galaxies, both spirals; the Triangulum galaxy (M33 or NGC 598) is a small spiral. The other galaxies are dwarf irregulars and ellipticals. There are 8 dwarf satellites orbiting M31, including the dwarf ellipticals M32 and NGC 205. There are 11 satellites of the Milky Way, including 9 dwarf spheroidals. One, the Sagittarius dwarf, was discovered as recently as 1994, in the direction of the Galactic Center about 24 kpc from the Sun. It is tidally elongated, with a mass $\sim 10^8$ $M_\odot$, similar to the other dwarfs. Several globular clusters appear to be associated with it. Metallicity constraints based on observations of the Milky Way's stellar halo indicate that fewer than 6 dwarf spheroidals have been accreted by the Milky Way in the last 10^{10} years.

The largest nearby dwarf galaxies are the Large and Small Magellanic Clouds, with masses of 2×10^{10} and 2×10^9 $M_\odot$, respectively. Orbital studies of the Magellanic Clouds indicate that they are only half as far away from us (about 55 kpc) as they were 10^{10} years ago and that they will merge with the Milky Way in another 10^{10} years.

12.6 Galaxy Clusters and Interactions

Galaxy clusters range from small groups of a few dozen members to large clusterings of thousands of galaxies; there are tens of thousands of clusters. Groups and clusters have space densities of galaxies that are 10 times greater than the density near them. The smaller clusters are called *open, diffuse*, or *irregular clusters* which have hundreds of members, sizes less then 10 Mpc, and a total mass of 10^{12}–10^{14} $M_\odot$. Larger clusters are called *rich* or *regular clusters*, with thousands of members, diameters up to 10 Mpc, and total mass of 10^{15} $M_\odot$. Open and rich clusters of galaxies have shapes similar to open and globular star clusters.

FIGURE 12.11 *Part of the Coma cluster. At the center is a cD galaxy, NGC 4881.
(Imaged with the Hubble Space Telescope WFPC2 by W. Baum, University of
Washington; the HST WFPC2 team, CalTech; and NASA. Image from the STScI public
website.)*

The former are loose and irregular; the latter have a central concentration of
galaxies. An example of a rich cluster is the Coma cluster, shown in Figure 12.11.

Diffuse galaxy clusters tend to have a higher proportion of spiral galaxies
than do the rich clusters, which contain mostly elliptical and S0 galaxies and often have cD galaxies in the center. Morphological distributions of galaxies in different clusters are shown in Figure 12.12.

This variation in morphology with cluster density suggests that the more fre-
quent interactions among dense cluster galaxies causes the shapes of galaxies to
change. For example, Hubble Deep Field observations indicate that the fraction
of lenticulars in distant clusters is lower than in closer ones. This result lends
support to the idea that lenticulars are made from spirals stripped of gas.

According to numerical simulations, two similar-mass spiral galaxies can
merge to form an elliptical galaxy. However, Hubble Deep Field observations
show that many distant ellipticals are red. They should be blue if all ellipticals
are formed by mergers, since colliding clouds of gas and dust would trigger
bursts of star formation that would be observable at high z. It is estimated that
perhaps 10% of elliptical galaxies were formed by mergers. Ellipticals formed in
this way are similar to galaxies that were born as ellipticals, but they have a
slightly extended outer component whose light falls off less steeply than $r^{1/4}$. The
centers of rich clusters tend to have giant cD elliptical galaxies. They are 10
times larger than normal ellipticals, perhaps because they have swallowed their
neighbors. The process is known as *galactic cannibalism*, in which a slow-moving
galaxy swings past the center of a dense cluster and merges with the central galaxy.

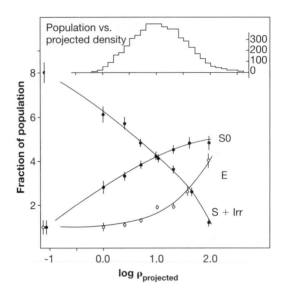

FIGURE 12.12 *Morphology of galaxies in clusters; the percentage of spirals decreases with increasing density (Dressler 1980). The top histogram shows the number of galaxies as a function of projected density for all clusters.*

One puzzle is that HST observations of the central regions of cDs generally reveal no distinct cores, even though calculations suggest that the nuclei of merged galaxies should have survived intact.

The process of ripping material from a close-passing galaxy is known as *tidal stripping* and occurs when the *Roche limit* is reached. This limit occurs where material from a galaxy experiences a stronger gravitational pull from another galaxy. The separation between galaxies at which this happens is the *Roche lobe*, of radius R:

$$R = \left(\frac{2M}{m}\right)^{1/3} r$$

where r is the radius of the small galaxy, and M and m are the masses of the large and small galaxy, respectively. Then, for example, a cD with 500 times the mass of its neighbor will tidal disrupt it where R = 10r.

Arm Classes are also correlated with environment: 2/3 of field spirals are flocculent, whereas an increasing fraction up to 2/3 of cluster spirals are grand design; the fraction increases with decreasing cluster crossing time. This result is consistent with the idea that neighbors can strip outer halos and generate density waves. Furthermore, interactions can induce bar formation, which also stimulates density waves.

Encounters with other galaxies can have a profound effect on a galaxy's morphology. Tidal interactions produce a number of distinctive features, such as tidal tails and bridges, spindles, rings, central ocular structures, and shells. They can also cause mass redistribution, leading to the formation of bars. Statistics indicate that they can shift a galaxy's Hubble type by 1/2 unit, for example from Sab to Sa, through mass transfer toward the center. The morphology depends on the types of galaxies interacting and on their relative masses, separations, and orientations. An example of a spindle galaxy, or polar ring galaxy, is shown in Figure 12.13. According to numerical simulations, a dwarf spiral galaxy that came too close was disrupted and now encircles the main galaxy.

A nearly head-on collision in which a small galaxy hits perpendicular to the plane of a spiral galaxy can produce a ring galaxy, as shown schematically in Figure 12.14. The interaction produces a "splash," which resulted in an outer ring where material collected. This was probably the fate suffered by the Cartwheel galaxy, shown from HST observations in Figure 12.15.

Differential gravitational effects disrupt material; in two nearby galaxies, the sides nearest each other feel the strongest gravitational pull, while the sides far-

FIGURE 12.13 *NGC 2685, the "spindle" galaxy, which has remnants of a merged galaxy wrapped around its central cigar shape. (Image from the STScI Digital Sky Survey.)*

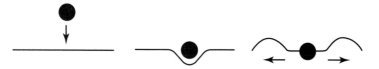

FIGURE 12.14 *Schematic side view of a head-on collision, in which a small galaxy hits the center of a spiral galaxy. The wave moves outward in the disk, collecting material into an outer ring at the wave crests.*

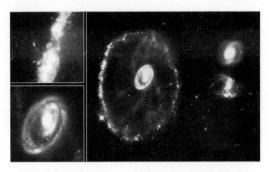

FIGURE 12.15   *(right) The Cartwheel galaxy. The spiral structure is beginning to reappear in the Cartwheel. One of the galaxies on the right may be the perturber. (top left) Enlargement of the ring, showing star formation clumps. (bottom left) Enlargement of the core, which contains young star clusters. (Imaged with Hubble Space Telescope WFPC2 by K. Borne, STScI; and NASA.)*

thest away feel the least pull. The centers feel an intermediate force. Thus, a bisymmetric tidal arm pattern is common, just as there are two tides on Earth from Earth-Moon tidal interactions (and also from Earth-Sun interactions). In Figure 12.16, two galaxies are separated by a distance R. Suppose the radius of one galaxy is r. For the near side, the gravitational force is

$$F_{near} = \frac{GMm}{(R - r)^2}$$

For the far side, the gravitational force is

$$F_{far} = \frac{GMm}{(R + r)^2}$$

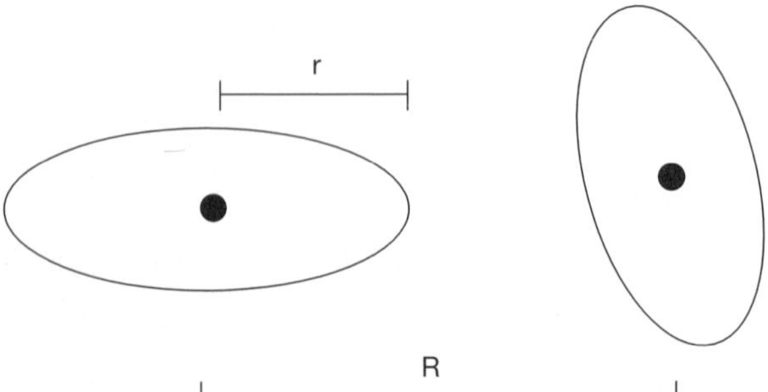

FIGURE 12.16   *Two nearby galaxies interact by tidal forces.*

Then the tidal force is given by the difference between the forces on the near and far side:

$$F_{tidal} \propto \frac{GMmr}{R^3}$$

Figure 12.17 shows tidal arms (sometimes called *tidal tails*) that resulted from near encounters between two galaxies. *Tidal bridges* of material between galaxies are also common. If two galaxies are oriented nearly perpendicular to each other, the galaxy moving perpendicular to the plane of another may create a warp in the disk rather than tidal arms. The detailed morphology depends also on the direction of rotation of a galaxy with respect to the path of its perturber; an encounter in the *prograde* direction (that is, with a path the same as the direction of rotation) has more pronounced effects than an encounter in the *retrograde* direction.

Numerical simulations of galaxy encounters have reproduced a wide variety of observed morphological features. A combination of optical and near-infrared surface brightness data and H I and CO radio observations provide parameters such as density contrasts and velocity fields that the models need to match. Supercomputers are now providing the capabilities of reproducing complex encounters, so that time scales, angles of encounter, and past and future morphological consequences can be investigated. Figure 12.18 shows an example of gas and stellar distributions in a simulation of two interacting galaxies.

Elliptical galaxies sometimes have low surface brightness shells of material surrounding them, as discovered in enhanced images by Arp, Malin, and Carter; see Figure 12.19. Computer simulations by F. Schweizer, L. Hernquist, P. Quinn, and others indicate that these shells are the result of elliptical galaxies assimilating smaller spiral galaxies in a head-on collision. The stars that have merged

FIGURE 12.17 *Interactions can produce tidal bridges, as shown here for NGC 5427 and 5428. (Image from the STScI Digital Sky Survey.)*

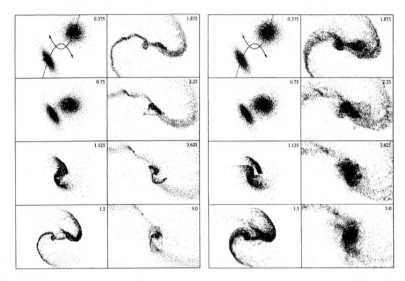

FIGURE 12.18   *Computer simulations showing the distributions of gas (left) and stars (right) as two galaxies swing past each other on parabolic orbits (indicated by arrows in the first frames). The frames increase in time starting from the top frame down. (From Barnes and Hernquist 1996.)*

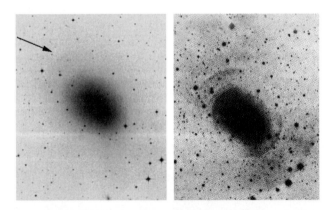

FIGURE 12.19   *NGC 3923 is an elliptical galaxy with many shells; the arrow points to a portion of one. (Left-hand image from the Digital Sky Survey; right-hand enhanced image from Malin and Carter 1983.)*

into the elliptical galaxy result in discrete shells because the stars spend more time in their orbits at apogalacticon than at perigalacticon, as shown in Figure 12.20. More than a dozen discrete shells may appear. Their spacing depends on the gravitational potential, which is inferred to be so large that dark matter must be present in the halos to account for it.

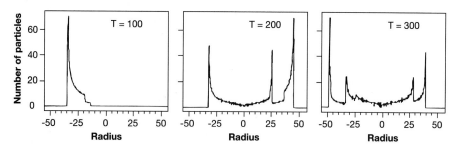

FIGURE 12.20 *Computer simulations show the number of particles as a function of distance from galaxy center as a result of an elliptical-spiral collision. Over three time intervals shown in the figure, the shells begin to appear as discrete collections of stars. (From Combes et al. 1995.)*

12.7 Superclusters

The Local Group is part of a *supercluster* of galaxies extending over a volume about 10 × 15 × 20 Mpc, as shown in Figure 12.21. The Virgo cluster is the largest cluster in this collection of clusters, so it is closest to the center of mass of the Local supercluster. It has about 205 members, 60% of which are spiral and 19% of which are elliptical. Since Virgo is visible from the Northern Hemisphere, it contains most of the nearby bright spirals that are commonly studied.

There are many surveys under way to explore the large-scale structure of the Universe. Some rely on 21-cm observations, others on optical redshifts. The most common type of survey involves taking a "slice" of the sky, a strip that is limited in declination and covers a large angle. An example is shown in Figure 12.22, which includes the Coma cluster. Large structures show up more readily in surveys of this type than in a single large area of the sky. The Universe consists of features that resemble a cross section of bubbles. The bubbles in this case have radii of tens of megaparsecs. The regions in the interiors of the bubbles are

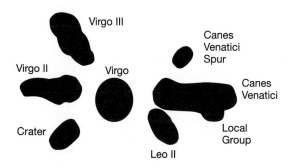

FIGURE 12.21 *Schematic diagram of the Virgo supercluster. (Based on Tully 1982.)*

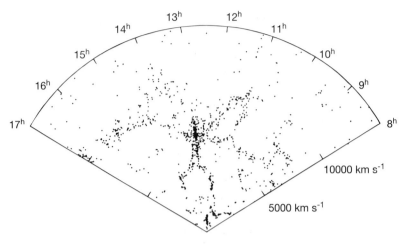

FIGURE 12.22   *Structure in the local Universe was mapped out from optical redshift surveys taking a slice of the sky. (From de Lapparent et al. 1986.)*

largely devoid of galaxies, so they are known as *voids*. The nearby Bootes void is 1 Mpc in diameter and does contain some dwarf galaxies. Nearby, seven discrete superclusters have been mapped. They have diameters of tens of megaparsecs and masses of $10^{15}–10^{17}$ M$_\text{o}$, corresponding to $10^4–10^6$ galaxies.

The so-called Seven Samurai are astronomers S. Faber, A. Dressler, D. Burstein, R. Davies, D. Lynden-Bell, R. Terlevich, and R. Wegner, whose studies of peculiar motions of galaxies in the Universe indicate that there is a strong gravitational attraction about 50 Mpc away. This region is known as the *Great Attractor*, which is a high concentration of matter nearly 150 Mpc long with a mass of about 5×10^{16} M$_\text{o}$, equivalent to a large supercluster. It is causing the Milky Way and other galaxies to stream toward it at a rate of 600 km s^{-1} in a direction orthogonal to the local Hubble flow. The Perseus-Pisces supercluster has about the same mass as the Great Attractor; they are the same distance from us but 180° apart from each other. A void separates us and the Perseus-Pisces supercluster, so we do not have as strong a gravitational pull in that direction as in the direction of the Great Attractor. Similarly, the *Great Wall* is a long string of galaxies spanning 230 Mpc. The wall appears to be the outer edge of a void and makes up part of the large-scale clumpiness of the Universe.

12.8 Galaxy Formation

Galaxies are believed to have formed out of perturbations that trace back to about 300,000 years after the Big Bang. This time is known as the *era of recombination* or the *era of decoupling*, since at this age the photons left over from the Big Bang had energies sufficiently low that electrons could reattach to nuclei to form atoms. The photons and matter were then distinct from each other, or decoupled. The average temperature of the Universe at decoupling was 3000 K. The cosmic background radiation photons have a redshift z = 1000. The ratio of the temperature of the Universe today to the temperature of the Universe at decoupling is approximately equal to the inverse redshift, so the average temperature of the Universe today is ~3 K (2.7 K).

Before decoupling, matter was ionized. It was linked to radiation by *Thompson scattering*, in which photons colliding with electrons transferred energy and increased their wavelength. Because matter and radiation were not separated during this time, radiation leaking out of a dense region dragged matter with it, thereby eliminating density fluctuations. After decoupling, perturbations larger than about 10^5 M_0 could grow.

Perturbations must have been present in the early Universe before decoupling in order for galaxies to form within 10^9 years after the Big Bang, as observed; otherwise, sufficiently dense fluctuations would not have had time to develop after decoupling. Evidence for such initial fluctuations was found with the COBE (Cosmic Background Explorer) satellite, in which microwave variations on the level of 1 part in 10^5 were observed (see Figure 12.23), or a range of ± 150 µK. These small fluctuations in the cosmic microwave background were discovered only after removing the microwave radiation from our own galaxy and accounting for the redshifted and blueshifted radiation from different parts of the sky due to our galaxy's random motion.

The relative densities of radiation and matter determine how perturbations behave. The mass density of radiation ρ_{rad} can be determined from the Stefan-Boltzmann law (energy flux = σT^4) and Einstein's $E = mc^2$:

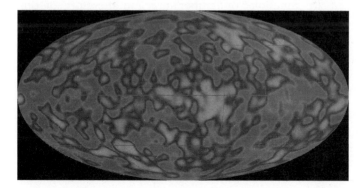

FIGURE 12.23 *COBE microwave fluctuations, with grid lines marking every 45°. The dipole radiation (motions of Earth, the Solar System, and the Milky Way) and Milky Way radiation have been subtracted. (Image from http://nssdca.gsfc.nasa.gov/anon_dir/cobe. The COBE datasets were developed by the NASA Goddard Space Flight Center under the guidance of the COBE Science Working Group and were provided by the NSSDC.)*

$$\rho_{\text{rad}} = \frac{4\sigma T^4}{c^3}$$

The mass density of radiation decreases as r^{-4} (since $T \propto 1/r$), while the mass density of matter decreases as r^{-3}. Therefore, at some point in time, the initially higher radiation density is exceeded by the matter density, as shown in Figure 12.24. This occurs at about 10^5 years after the Big Bang, when the Universe is said to go from being *radiation-dominated* to being *matter-dominated*. The exact age at which the two densities are equal depends on the deceleration parameter and the Hubble constant.

In order to understand the growth of structure in the early Universe, we will assume that the Jeans gravitational instability is the primary process of density enhancements. Recall from Chapter 10 that the Jeans length for gravitational instabilities increases with increasing velocity dispersion in a disk. In the expanding Universe, the Jeans length depends on the sound speed a, and there is a relativistic effect at early times because the local energy density is high. In that case, the density ρ is replaced by $\rho + (P/c^2)$ for pressure P. Then the Jeans mass for a three-dimensional collapse may be written as

$$M_J = \left(\frac{4\pi mn}{3}\right)\left(\frac{\pi a^2}{G\left(\rho + \dfrac{P}{c^2}\right)}\right)^{3/2}$$

for hydrogen mass m and total baryon number density n. The pressure and sound speed are functions of temperature and therefore of time. Before

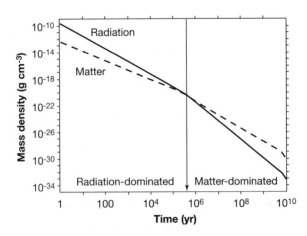

FIGURE 12.24 *Mass densities of radiation and matter in the Universe.*

decoupling, the sound speed was close to the speed of light, and the corresponding Jeans mass was ~10^{17} $M_\odot$. After decoupling, the sound speed decreased to a few kilometers per second and the Jeans mass fell to a few times 10^5 $M_\odot$ (a factor of 10^{12} lower). The Jeans mass as a function of time in an expanding Universe is shown schematically in Figure 12.25; the abrupt drop occurs at decoupling.

Suppose that there are density perturbations in the early Universe that have masses greater than the Jeans mass. They will grow with time until the Universe's temperature from expansion is sufficiently low that their mass is just equal to the Jeans mass. Their growth can be described in terms of a wave equation, such as we have seen in the dispersion relations in Appendix 4. In Figure 12.25, it is evident that a perturbation with a mass of 10^{12} $M_\odot$ is gravitationally unstable to growth until about 10 years after the Big Bang, after which its mass is less than the Jeans mass and further growth is halted. After decoupling, that perturbation may grow again as the radiation pressure drops and the Jeans mass decreases. Small perturbations are damped, and the boundary between growth and damping is marked "damping" in Figure 12.25. The resultant masses may be the size of globular clusters or of clusters of galaxies.

Perturbations in general may either be adiabatic or isothermal. Adiabatic fluctuations are disturbances in which no heat is exchanged between an object and its surroundings. In this case, the equation of state is given by $P \propto \rho^\gamma$, where γ is

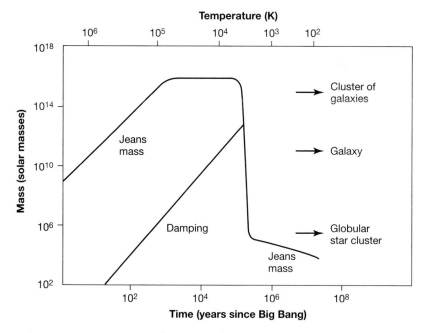

FIGURE 12.25 *Preferred scale sizes in the early Universe. (Adapted from Silk 1989.)*

the ratio of specific heats at constant pressure and constant volume; $\gamma = 5/3$ for a monatomic gas. In the early Universe, adiabatic fluctuations would cause compression of both matter and radiation, and no radiation could escape. These fluctuations would be damped out before decoupling and would be gravitationally unstable afterward on scales larger than 10^{13} $M_\odot$. In isothermal fluctuations, the temperature of the object remains constant; then the equation of state is $P \propto \rho$ for an ideal gas; in the early Universe, these fluctuations would correspond to perturbations only in the matter density and not in the radiation density. Such fluctuations would not grow until after decoupling and then would be gravitationally unstable if their masses exceeded $\sim 10^6$ $M_\odot$.

There are different theories about how galaxies formed in the billion years after the Big Bang. Zel'dovich popularized a *top-down scenario*, in which units with the mass of galaxy clusters, $\sim 10^{17}$ $M_\odot$, formed first and then fragmented to form individual galaxies. This adiabatic model is sometimes referred to as the *pancake model*, because unstable masses are rarely spherical so they tend to collapse faster along one axis than another. In contrast, J. Peebles proposed a *bottom-up scenario*. Here, the preferred unit formed by the Universe was $\sim 10^5$ $M_\odot$, the mass of globular clusters, which then combined to form galaxies.

These two scenarios are linked to cold dark matter (CDM) and hot dark matter (HDM) (as discussed in Chapter 7). In CDM models, density perturbations begin to grow at decoupling. Cold dark matter does not interact with baryonic matter except by gravity, so radiation escaping from dense regions does not alter the density. In addition, cold dark matter has no thermal pressure to resist gravitational collapse. In HDM models, neutrinos eliminate all but the largest-scale fluctuations because they are so energetic that they escape dense regions and smooth out small fluctuations.

Although a detailed discussion of the very early Universe is beyond the scope of this book, we note that inflationary models are consistent with CDM models, whereas cosmic string models are connected with HDM models. *Inflation* is an event that cosmologists believe may have occurred in a period $t \approx 10^{-35}$ s to perhaps $t \approx 10^{-30}$ s after the Big Bang. During inflation, the Universe doubled its size about 100 times due to the release of energy from the *phase transition* when the strong force separated from the other forces. *Cosmic strings* are regions where the four forces of nature did not become distinct; they would be extremely thin but have very high densities (10^{22} g cm^{-3}). The strings would act as seeds for the largest-scale structures observed in the Universe, either by attracting matter or by blowing out holes (cosmic voids) from vibrations.

One difficulty with HDM models and the top-down scenario is that individual galaxies would have formed out of the most recent of the fragmentations. Consequently, we should only see galaxies at $z < 1$ (where indeed most galaxies reside). However, galaxies are also observed at redshifts greater than 3, quasars at $z > 4$, and protogalaxies at $z > 6$, which is not expected in the top-down scenario. HDM models do not adequately produce the smallest-scale structures and also produce large-scale structure that is more prominent than what is observed.

On the other hand, CDM models do not completely account for the largest-scale structures. Some astronomers believe that a mixture of CDM and HDM can provide the best match of growth time scales and large-scale structures; a model with approximately 2/3 CDM fits the large- and small-scale observations, with 1/3 HDM to smooth out the smallest scales. The results of a mixed-model n-body simulation are shown in Figure 12.26.

Figure 12.27 shows a gravitational lens formed by distant galaxies passing through the line of sight of a closer cluster of galaxies.

Quasars are often surrounded by discrete clouds, which are observed as Lyman α absorption features at different redshifts. There are so many absorption lines that they are often referred to as the *Lyman α forest*. These clouds are evidently galaxies in the process of forming. A Key Project of the Hubble Space Telescope is to take spectra of these absorption line systems in order to understand their motions with respect to the quasars. Figure 12.28 is the Hubble deep field with spectra from real quasars superposed; the ovals represent primordial clouds around these galaxies, and the lines indicate how the Lyman α forest would differ for different observed lines of sight.

Our Universe of galaxies is complex and dynamic. Just as some puzzles are solved about the evolutionary history and future of galaxies, new mysteries invariably appear. The next decade should witness considerable progress in the field of galaxy research as the Hubble Space Telescope and new generation

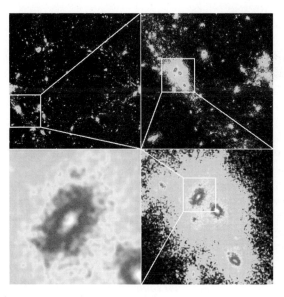

FIGURE 12.26   *Mixed dark matter model, showing scales of 100 Mpc (top left), 24 Mpc (top right), 6 Mpc (bottom left), and 1.5 Mpc (bottom right). (By C.P. Ma and E. Bertschinger, Massachusetts Institute of Technology.)*

FIGURE 12.27 *Gravitational lens from Abell cluster 2218. The Abell cluster is on the line of sight to distant faint blue galaxies, which are distorted into arcs as their light passes through the cluster. (Imaged with Hubble Space Telescope WFPC2 by W. Couch, University of New South Wales; R. Ellis, Cambridge University; and NASA.)*

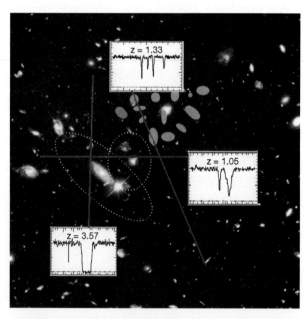

FIGURE 12.28 *Schematic diagram showing how absorption lines along the lines of sight to distant quasars indicate the presence of clouds that may be protogalaxies. (From C. Churchill and J. Charlton 1997, website http://www.astro.psu.edu/users/charlton.)*

ground-based telescopes and supercomputers assist in our quest to gather and interpret observations and refine theories.

Exercises

1. Starting with the equation for gravitational force, show why the tidal force has an inverse-cube dependence on distance instead of an inverse-square dependence.
2. On the basis of the Schechter luminosity function, how many more galaxies per unit volume are expected with absolute magnitudes of -20 than with absolute magnitudes of -22?
3. If a Cepheid observed in M81 with the Hubble Space Telescope has an apparent magnitude of $m_V = 25.6$ and a period of 5 days, determine the distance to the galaxy.
4. Suppose a globular cluster luminosity function is observed in another galaxy with points (Φ per mag, apparent magnitude) = (1, 15), (6, 16), (32, 17), (79, 18), (126, 19), (100, 20), (63, 21), ((37, 22). What is the distance of the galaxy?
5. For a current density $\rho_0 = 10^{-31}$ g cm^{-3}, determine the value of the Hubble constant that would correspond to a flat universe.
6. NGC 4321 has six dwarf companions. Locate them on the POSS or from the DSS. Assume that their masses scale with their sizes. On the basis of their apparent separations from NGC 4321, determine the relative strengths of their tidal interactions.
7. Using NED or another database, determine the relative tidal forces of the Large and Small Magellanic Clouds on the Milky Way by estimating their masses and separations from us.

Unsolved Problems

1. What is the value of the Hubble constant and the age of the Universe?
2. How do galaxies form?

Useful Websites

"A Digital Sky Survey of the Northern Galactic Cap" by the Sloan Digital Sky Survey research group at Princeton University is available via
http://www.astro.princeton.edu/GBOOK/

At the Shodor Education Foundation website, students can make use of a web-based interface to interactive simulations, including galaxy formation. Software can also be downloaded to create similar models on personal computers, all via
http://www.shodor.org/MASTER/galaxy_home.html

The full Las Campanas Redshift Survey is available online:
http://www.manaslu.astro.utoronto.ca/~lin/catalog/catalog/html

Cosmological structure simulations are on http://zebu.uoregon.edu/cosmos.edu and
http://zebu.uoregon.edu/movie.html
http://arcturus.mit.edu/gallery/galaxy.html

http://www.kusmos.phsx.ukans.edu
http://www.cnde.iastate.edu/staff/cosmology.html

Hubble Deep Field images and spectra are available at
http://astron.berkeley.edu/davisgrp/HDF/observations.html and
http://www.stsci.edu

Quasar absorption line research is on http://www.astro.psu.edu

Further Reading

Arnaboldi, M., T. Oosterloo, F. Combes, K. Freeman, and B. Koribalski. 1997. New H I observations of the prototype polar ring galaxy NGC 4650A. *Astronomical Journal* 113:585.

Bahcall, N. 1988. Large-scale structure in the universe indicated by galaxy clusters. *Annual Review of Astronomy and Astrophysics* 26:631.

Barnes, J., and L. Hernquist. 1992. Dynamics of interacting galaxies. *Annual Review of Astronomy and Astrophysics* 30:705.

Barnes, J., and L. Hernquist. 1996. Transformations of galaxies. II. Gasdynamics in merging disk galaxies. *Astrophysical Journal* 471:115.

Bartusiak, M. 1997. What makes galaxies change? *Astronomy* 25:36.

Binggeli, B., A. Sandage, and G. Tammann. 1988. The luminosity function of galaxies. *Annual Review of Astronomy and Astrophysics* 26:509.

Bonometto, S., A. Iovino, L. Guzzo, R. Giovanelli, and M. Haynes. 1993. Correlation functions from the Perseus-Pisces redshift survey. *Astrophysical Journal* 419:451.

Burstein, D., S. Faber, and A. Dressler. 1990. Evidence from the motions of galaxies for a large-scale, large-amplitude flow toward the great attractor. *Astrophysical Journal* 354:18.

Butcher, H., and A. Oemler. 1978. The evolution of galaxies in clusters. I. ISIT photometry of CL00211+1654 and 3C295. *Astrophysical Journal* 219:18.

Combes, F., P. Boisse, A. Mazure, and A. Blanchard. 1995. *Galaxies and cosmology*, trans. M. Seymour. New York: Springer.

Contopoulos, G., and D. Kotsakis. 1987. *Cosmology.* New York: Springer-Verlag.

DaCosta, L., W. Freudling, G. Wegner, R. Giovanelli, M. Haynes, and J. Salzer. 1996. The mass distribution in the nearby universe. *Astrophysical Journal Letters* 468:5.

de Lapparent, V., M. Geller, and J. Huchra. 1986. A slice of the universe. *Astrophysical Journal Letters* 302:L1.

Dickey, J.M. 1988. *The Minnesota Lectures on Clusters of Galaxies and Large-Scale Structure*: Astronomical Society of the Pacific Conference Series, vol. 5. San Francisco: Astronomical Society of the Pacific.

Dressler, A. 1980. Galaxy morphology in rich clusters. Implications for the formation and evolution of clusters. *Astrophysical Journal* 236:351.

Dressler, A. 1984. The evolution of galaxies in clusters. *Annual Review of Astronomy and Astrophysics* 22:185.

Dressler, A., A. Oemler, H. Butcher, and J. Gunn. 1994. The morphology of distant cluster galaxies. I. HST Observations of CL0939+4713. *Astrophysical Journal* 430:107.

Elmegreen, D.M., B. G. Elmegreen, and A. D. Bellin. 1990. Statistical evidence that galaxy companions trigger bars and change the Hubble type. *Astrophysical Journal* 364:415.

Fabian, A. 1994. Cooling flows in clusters of galaxies. *Annual Review of Astronomy and Astrophysics* 32:277.

Feast, M., and A. Walker. 1987. Cepheids as distance indicators. *Annual Review of Astronomy and Astrophysics* 25:345.

Felten, J. 1977. Study of the luminosity function of field galaxies. *Astronomical Journal* 82:861.

Ferguson, H. 1992. The galaxy luminosity function in different environments. In *Physics of nearby galaxies: Nature or Nurture?: XIIth Moriond Astrophysics Meeting*, ed. T. Thuan, C. Balkowski, and J. Tran Thanh Van, 443. Paris, France: Editions Frontieres.

Ferguson, H., R. Williams, and L. Cowie. 1997. Probing the faintest galaxies. *Physics Today* 50:24.

Freedman, W., et al. 1994. The Hubble Space Telescope extragalactic distance scale key project. 1: The discovery of Cepheids and a new distance to M81. *Astrophysical Journal* 427:628.

Freudling, W., L. DaCosta, G. Wegner, R. Giovanelli, M. Haynes, and J. Salzer. 1995. Determination of Malmquist bias and selection effects from Monte Carlo simulations. *Astronomical Journal* 110:920.

Gaidos, E. 1997. The galaxy luminosity function from observations of twenty Abell clusters. *Astronomical Journal* 113:117.

Ghigna, S., S. Borgani, S. Bonometto, L. Guzzo, A. Klypin, J. Primack, R. Giovanelli, and M. Haynes. 1994. Sizes of voids as a test for dark matter models. *Astrophysical Journal Letters* 437:71.

Giovanelli, R., and M. Haynes. 1991. Redshift surveys of galaxies. *Annual Review of Astronomy and Astrophysics* 29:499.

Harris, W., J. Allwright, C. Pritchet, and S. van den Bergh. 1991. The luminosity distribution of globular clusters in three giant Virgo ellipticals. *Astrophysical Journal (Suppl.)* 76:115.

Haynes, M., and R. Giovanelli. 1988. Large-scale structure in the local universe: The Pisces-Perseus supercluster. In *Large-scale motions in the universe: A Vatican study week*, ed. V. Rubin and G. Coyne, 31. Princeton: Princeton University Press.

Haynes, M., and R. Giovanelli. 1991. Neutral hydrogen observations of galaxies in superclusters. *Astrophysical Journal (Suppl.)* 77:331.

Haynes, M., R. Giovanelli, and G. Chincarini. 1984. The influence of environment on the H I content of galaxies. *Annual Review of Astronomy and Astrophysics* 22:445.

Hernquist, L., and P. Quinn. 1988. Formation of shell galaxies. I. Spherical potentials. *Astrophysical Journal* 331:682.

Hoffman, G., E. Salpeter, and G. Helou. 1990. Blue compact dwarf galaxies in Virgo and nearby groups. In *Astronomical Society of the Pacific Conference Series 10: Evolution of the universe*, ed. R. Kron, 67. San Francisco: Astronomical Society of the Pacific.

Huchra, J., M. Geller, V. de Lapparent, and H. Corwin. 1990. The CFA redshift survey—data for the NGP+30 zone. *Astrophysical Journal (Suppl.)* 72:433.

Humphreys, R., and A. Sandage. 1980. On the stellar content and structure of the spiral galaxy M33. *Astrophysical Journal (Suppl.)* 44:319.

Ibata, R., R. Wyse, G. Gilmore, M. Irwin, and N. Suntzeff. 1997. The kinematics, orbit, and survival of the Sagittarius dwarf spheroidal galaxy. *Astronomical Journal* 113:634.

Impey, C., G. Bothun, and D. Malin. 1988. Virgo dwarfs: New light on faint galaxies. *Astrophysical Journal* 330:634.

Iovino, A., R. Giovanelli, M. Haynes, G. Chincarini, and L. Guzzo. 1993. Galaxy clustering, morphology, and luminosity. *Monthly Notices of the Royal Astronomical Society* 265:21.

Jacoby, G., D. Branch, R. Ciardullo, R. Davies, W. Harris, M. Pierce, C. Pritchet, J. Tonry, and D. Welch. 1992. A critical review of selected techniques for measuring extragalactic distances. *Publications of the Astronomical Society of the Pacific* 104:599.

Kennicutt, R., W. Freedman, and J. Mould. 1995. Measuring the Hubble constant with the Hubble Space Telescope. *Astronomical Journal* 110:1476.

Koo, D., and R. Kron. 1992. Evidence for evolution in faint field galaxy samples. *Annual Review of Astronomy and Astrophysics* 30:613.

Malin, D., and D. Carter. 1983. A catalog of elliptical galaxies with shells. *Astrophysical Journal* 274:534.

Narlikar, J., and T. Padmanabhan. 1991. Inflation for astronomers. *Annual Reviews of Astronomy and Astrophysics* 29:325.

Norman, C. 1988. Quasar absorption lines and galaxy formation. In *Confrontations between Theory and Observations in Cosmology*, ed. J. Audouze and F. Melchiorri. Cambridge: Cambridge University Press.

Peebles, J. 1996. The primeval mass fluctuation spectrum and the distribution of nearby galaxies. *Astrophysical Journal* 473:42.

Roberts, M. S., and M. P. Haynes. 1994. Physical parameters along the Hubble sequence. *Annual Review of Astronomy and Astrophysics* 32:115.

Salzer, J.J. 1989. Observations of a complete sample of emission-line galaxies. III. Spatial and luminosity distributions of the UM galaxies. *Astrophysical Journal* 347:152.

Sandage, A., and G. Tammann. 1995. Steps toward the Hubble Constant. X. The distance of the Virgo Cluster core using globular clusters. *Astrophysical Journal* 446:1.

Schechter, P. 1976. An analytic expression for the luminosity function for galaxies. *Astrophysical Journal* 203:297.

Schectman, S., S. Landy, A. Oemler, D. Tucker, H. Lin, R. Kirshner, and P. Schechter. 1996. The Las Campanas redshift survey. *Astrophysical Journal* 470:172.

Schweizer, F., and P. Seitzer. 1988. Ripples in disk galaxies. *Astrophysical Journal* 328:88.

Seldner, M., B. Siebers, E.J. Groth, and P.J.E. Peebles. 1977. New reduction of the Lick catalog of galaxies. *Astronomical Journal* 82:249.

Silbermann, N., et al. 1996. The Hubble Space Telescope key project on the extragalactic distance scale. VI. The Cepheids in NGC 925. *Astrophysical Journal* 470:1.

Silk, J. 1989. *The Big Bang.* New York: W.H. Freeman.

Stengler-Larrea, E., et al. 1995. The Hubble Space Telescope quasar absorption line key project. V. Redshift evolution of Lyman limit absorption in the spectra of a large sample of quasars. *Astrophysical Journal* 444:64.

Struck, C., P. Appleton, K. Borne, and R. Lucas. 1996. Hubble Space Telescope imaging of dust lanes and cometary structures in the inner disk of the Cartwheel ring galaxy. *Astronomical Journal* 112:1868.

Thronson, H., and J.M. Shull. 1990. *Second Wyoming Conference: The interstellar medium in galaxies.* Dordrecht: Kluwer Academic.

Tully, R.B. 1982. Unscrambling the local supercluster. *Sky and Telescope* 63:550.

van den Bergh, S., and C. Pritchet. 1988. *The extragalactic distance scale: Proceedings of the PASP 100th Anniversary Symposium, Astronomical Society of the Pacific Conference Series,* vol. 4. San Francisco: Astronomical Society of the Pacific.

White, M., D. Scott, and J. Silk. 1994. Anisotropies in the cosmic microwave background. *Annual Review of Astronomy and Astrophysics* 32:319.

Williams, R., et al. 1996. The Hubble deep field: Observations, data reduction, and galaxy photometry. *Astronomical Journal* 112:1135.

Zeldovich, Y. 1982. The origin of large-scale cell structure in the Universe. *Soviet Astronomy Letters* 8:102.

APPENDIX 1

Constants

Physical constant	Symbol	Value
Speed of light	c	2.998×10^{10} cm s^{-1}
Gravitational constant	G	6.668×10^{-8} dyn cm^2 g^{-2}
Planck constant	h	6.625×10^{-27} erg s
Boltzmann constant	k	1.380×10^{-16} erg K^{-1}
Density of water	ρ	1 g cm^{-3}
Rydberg constant	R	1.097×10^5 cm^{-1}
Stefan-Boltzmann constant	σ	5.670×10^{-5} erg K^{-4} cm^{-2} s^{-1}
Electron mass	m_e	9.110×10^{-28} g
Proton mass	m_p	1.673×10^{-24} g
Electron charge	e	4.803×10^{-10} esu
Wien displacement constant	C	2.90×10^{-1} cm K
Recombination coefficient for H	α	6.82×10^{-13} cm^3 s^{-1} @ 5000 K
		4.18×10^{-13} cm^3 s^{-1} @ 10000 K
Hydrogen ionization	H$^+$	13.6 eV ionization energy
Helium ionization	He$^+$	54.4 eV ionization energy

APPENDIX 2

Astronomical Quantities

Quantity	Abbreviation	Value
Parsec	pc	3.086×10^{18} cm
Megaparsec	Mpc	10^6 pc $= 3.086 \times 10^{19}$ km
Astronomical unit	AU	1.496×10^{13} cm
Solar mass	M_o	1.989×10^{33} g $= 1.989 \times 10^{30}$ kg
Solar luminosity	L_o	3.826×10^{33} erg s^{-1}
Solar radius	R_o	6.960×10^{10} cm
Sun's distance from Galactic Center	R_o	8.5 kpc (IAU value)
Sun's orbital speed	V_o	220 km s^{-1} (IAU value)
Hubble constant	H_o	100h km s^{-1} Mpc^{-1}, for h between 0.5 and 1
Critical density	ρ_c	$3 H_o^2/(8\pi G) = 1.1 \times 10^{-29}$ g cm^{-3} for $H_o = 75$
Ratio of density to critical density	Ω	0.1 to 1 (uncertain)
Ratio of total to selective extinction	R	$A_V/E(B\text{-}V) = 3.2$ (varies)
Oort's constants	A	14.5 ± 1.5 km s^{-1} kpc^{-1}
	B	-12 ± 3 km s^{-1} kpc^{-1}

APPENDIX 3
Conversions

Quantity	Conversion
Velocity	1 km/s = 1 pc/10^6 yr (approximately)
Time	1 yr = 3.16×10^7 s
Flux unit	1 Jy = 10^{-26} W m^{-2} Hz^{-1}
Power	watt = joule s^{-1}
Energy	joule = 10^7 erg s^{-1}
	1 eV = 1.602×10^{-12} erg
Density	1 g cm^{-3} = 10^3 kg m^{-3}
Wavelength	1 Å = 10^{-8} cm = 0.1 nm = $10^{-4} \mu$
Gravitational constant	dyn cm^2 g^{-2} = erg cm g^{-1} = 10^{-3} N m^2 kg^{-2}
Distance	1 pc = 3.262 light years
	1 light year = 6.324×10^4 AU
	1 AU = 1.496×10^{13} cm = 1.496×10^8 km

APPENDIX 4

Fluid Dynamics

A4.1 Basic Hydrodynamic Equations

In Chapters 9 and 10 we considered spiral density waves and star formation. Here we examine the theory behind waves and gravitational instabilities in more detail. The equations of *momentum conservation, gravitational potential* (Poisson's equation), and *continuity* are used to derive an equation for a stellar fluid that relates ω_p, ω, and k; this is called a *dispersion relation*. Let us briefly review these fundamental hydrodynamic equations.

The equation of momentum conservation can be expressed in many ways for fluids. Fluid pressure (P) and gravity (potential Φ) are related to the density (ρ) of the medium and its velocity flow (v). For macroscopic motions, we can write:

$$\rho \frac{Dv}{Dt} \equiv \rho \frac{\partial v}{\partial t} + \rho\,(v\cdot\nabla v) = -\nabla P - \rho\nabla\Phi$$

where the left-hand term has a Lagrangian time derivative, which is comoving with the fluid, and the middle term with partial derivatives expresses the Eulerian time derivative, which is the motion viewed from a fixed location outside the fluid. The pressure term is a given by $P = \rho a^2$ for one-dimensional velocity dispersion a.

Poisson's equation expresses the gravitational potential at any time in terms of the density:

$$\nabla^2\Phi = 4\pi\,G\rho$$

It is a second-order partial differential equation. In a thin spiral disk, the three-dimensional density ρ is replaced with a two-dimensional density σ (gm cm^{-2}), also called the *surface density* or *mass column density*.

295

The equation of continuity says that as fluid particles move from one place to another, they are neither created nor destroyed. The density changes as the fluid diverges and converges, such that:

$$\frac{D\rho}{Dt} + \rho(\nabla \cdot v) = \frac{\partial \rho}{\partial t} + \nabla \cdot (\rho v) = 0$$

A4.2 Dispersion Relation

In a rotating disk with tightly wound spirals, these equations can be rewritten and combined to form the following *dispersion relation* for the spiral wave:

$$m^2(\omega - \omega_p)^2 = -2\pi G\sigma_0|k| + k^2a^2 + \kappa^2$$

The term $2\pi G\sigma_0|k|$ is a gravity term, and k^2a^2 is a pressure term. As before, κ is the epicyclic frequency, and a is the velocity dispersion for the stars; k is a wave number ($= 2\pi/\lambda$) that gives the spacing between arms. The initial surface density is σ_0. The pressure term and the differential rotation term (κ^2) are stabilizing forces in the disk, which means that they help sustain oscillatory disturbances. The gravity term is destabilizing since it has a negative sign associated with it; that means there are exponentially growing disturbances.

The term $m(\omega - \omega_p)$ is called the *relative driving frequency*; the epicyclic frequency is sometimes called the *natural frequency* of the system. The resonances come from this equation. In between the OLR and ILR, the relative driving frequency is less than the natural frequency κ, so waves are sustained. Outside these limits, the relative driving frequency is too large, and self-sustained spiral density waves are not possible.

There are different solutions for the waves in the dispersion relation depending on whether the gravity or pressure terms dominate and whether k is positive or negative. We can solve the dispersion relation for k. For convenience, let $Q = a\kappa/\pi G\sigma_0$; this is called the *Toomre parameter* for stability, as described in Chapter 10. From the quadratic formula, we have:

$$k = \pm \frac{\pi G\sigma_0}{a^2}\left[1 \pm \sqrt{\left(1 - Q^2 + \frac{m^2(\omega - \omega_p)^2}{\kappa^2}Q^2\right)}\right]$$

When k is large (corresponding to a plus sign outside the square root term), pressure dominates because it scales with k^2. The wavelength scales inversely with k, so the solution in this case is said to be a *short-wavelength solution*. When k is small (corresponding to the minus sign in front of the square root term), the pressure term is small too, so the gravity term dominates. In this case, the solution is called a *long-wavelength solution*. In addition, waves may be leading or trailing, depending on the sign of k: For a galaxy rotating in the direction of increasing angle θ, for k > 0, the waves are trailing; for k < 0, the waves are leading. There can be solutions of any combination: short trailing or leading waves, long trailing or leading waves. The overall wave motion is such that short, trailing waves move away from corotation: Inside corotation, they move toward the center; outside corotation, they move outward. Long leading waves behave the same as short trailing waves, but long trailing or short leading waves move in the opposite direction. The short, trailing solution is the optical manifestation that we see as the main spiral arms. Table A4.1 summarizes these solutions.

Short trailing waves spiral to the center with a group velocity given by $\partial \omega_p / \partial k$. The time scale is ~$2 \times 10^8$ yr to travel across the disk (i.e., about the same as the rotation period), so spiral density waves do not last long unless they are reflected and reinforced.

Calculations of the small perturbations in mass column density σ about the average disk value caused by a linear wave show that, if the potential of the arm follows a $-\cos(m\omega_p t - m\theta)$ variation for phase θ and number of arms m, then the mass column density enhancement, σ_1, follows a cosine variation too, $\cos(m\omega_p t - m\theta)$ but with positive cosine, as shown in Figure A4.1. That is, the density varies around the galaxy, and the density peaks, identified as the spiral arms, occur at the potential minima. The phase, θ, tells where the arms are in azimuthal angle, as shown in Figure 5.14. The wave number, k, can be written in terms of the pitch angle i: k = 2/[r tan i] = radial derivative of phase θ, where i > 0 means the arms are trailing.

The radial and azimuthal components of motion are given by $v_r \propto -\cos(m\omega_p t - m\theta)$ and $v_\theta \propto \sin(m\omega_p t - m\theta)$; that is, the velocity components vary as cosines and sines. In the middle of the arm, v_r is a negative maximum in the region where the density enhancement σ_1 is a maximum, and v_θ, the azimuthal component, is 0.

In Section 10.2, we discussed the Jeans length and growth rate of clouds; these are more precisely derived by using a dispersion relation such as for spiral density wave perturbations. A dispersion relation for disk instabilities similarly

TABLE A4.1 *Solutions to the dispersion relation for a density wave*

Sign of k	Wave orientation	Size of k	Wavelength	Wave motion
+	Trailing	Large	Short	Away from corotation
−	Leading	Small	Long	Away from corotation
−	Leading	Large	Short	Towards corotation
+	Trailing	Small	Long	Towards corotation

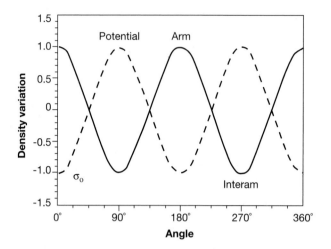

FIGURE A4.1 *Density variation is shown as a function of phase angle for a spiral density wave with two arms.*

requires an equation of motion, an energy equation, and a periodic perturbation. Then the dispersion relation in a two-dimensional disk is

$$\omega^2 = -k^2 a^2 + 2\pi G\Sigma k - \kappa^2$$

The wave number k for peak growth occurs at

$$k = \frac{\pi G\Sigma}{a^2}$$

so the equivalent Jeans length is

$$\lambda = \frac{2\pi}{k} = \frac{2a^2}{G\Sigma}$$

The growth rate at this wave number is determined from a quadratic solution, which gives:

$$\omega = \frac{\pi G\Sigma}{a}$$

and the time scale for collapse is the inverse of this frequency, or

$$t = \frac{a}{\pi G\Sigma}$$

A4.3 Shocks

Shocks can arise when fast-moving material runs into slower-moving material. The local sound speed a of a gas is given by

$$a = \sqrt{\frac{\gamma P}{\rho}}$$

for gas pressure P, density ρ, and ratio of specific heats γ. The value of γ is 5/3, for an *adiabatic atomic gas*, or 1 for an *isothermal gas*. An adiabatic gas is one in which there is no radiation; this happens, for example, when there is a perturbation that is fast compared with the radiative cooling time. The temperature T of a gas is given by P = nkT for number density n. The pressure is related to density by $P \propto \rho^{\gamma}$, so $T \propto \rho^{\gamma-1}$. When a gas with $\gamma > 1$ is compressed, it heats up and its sound speed increases. Then the hot compressed gas tries to overtake the cooler uncompressed gas in front of it, which results in a *discontinuity* at the boundary of the two regions. This boundary is a shock front. The density of the compressed gas ρ_{post} compared with the uncompressed gas ρ_{pre} is given by

$$\frac{\rho_{pre}}{\rho_{post}} = \frac{3}{4}M^{-2} + \frac{1}{4}$$

where M is the *Mach number* of the gas, defined by the ratio of the speed of the shock front to the sound speed of the unshocked gas. The maximum compression possible in a strong adiabatic shock, where $M >> 1$, is 4, so high compression does not occur.

In an isothermal gas, energy gained during compression is radiated away, so the temperature does not increase. Then the maximum compression is given by

$$\frac{\rho_{post}}{\rho_{pre}} = M^2$$

This compression can be much higher than the maximum achieved in an adiabatic shock; a factor of 100 is not uncommon in the interstellar medium. Therefore, shocks can lead to the formation of molecular clouds from atomic clouds or to star formation in pre-existing molecular clouds, provided other conditions are satisfied.

Further details about dispersion relations, shock fronts, and boundary conditions can be found in the references listed below.

Further Reading

Elmegreen, B. G. 1992. Large-scale dynamics of the interstellar medium. In *The galactic interstellar medium: Saas-Fee advanced course 21*, ed. D. Pfenniger and P. Bartholdi, 157. New York: Springer-Verlag.

Scheffler, H., and H. Elsässer. 1987. *Physics of the galaxy and interstellar matter*. New York: Springer-Verlag.

Toomre, A. 1977. Theories of spiral structure. *Annual review of Astronomy and Astrophysics* 15: 437.

Index